PERSPECTIVES IN HUMAN REPRODUCTION

E. S. E. HAFEZ
Editor-in-Chief

Volume 4
SCANNING ELECTRON MICROSCOPY OF HUMAN REPRODUCTION

Edited by

E. S. E. HAFEZ

Reproductive Physiology Laboratories
C. S. Mott Center for Human Growth and Development
Wayne State University School of Medicine
Detroit, Michigan

ANN ARBOR SCIENCE
PUBLISHERS INC
P.O. BOX 1425 • ANN ARBOR, MICH. 48106

Copyright © 1978 by Ann Arbor Science Publishers, Inc.
230 Collingwood, P. O. Box 1425, Ann Arbor, Michigan 48106

Library of Congress Catalog Card No. 77-85087
ISBN 0-250-40181-9

Manufactured in the United States of America
All Rights Reserved

PREFACE

Scanning electron microscopy became available commercially in late 1965, and by 1969 the technique was being applied extensively to biomedical research. This technique has provided dramatic new information which has led to the reinterpretation of many biological structures that had previously been observed only superficially by conventional light microscopy. Studies using transmission electron microscopy are time-consuming, and some of the observations may be difficult to interpret. Scanning electron microscopy permits rapid visualization of larger surfaces of tissues, and the preparation is relatively rapid.

Scanning electron microscopy overcomes the problems associated with restricted magnification and depth of focus in light microscopy, and those associated with the limited dimensionality of transmission electron microscopy. Precise examinations become possible for large surface areas and detailed surface ultrastructure in specifically selected regions.

A workshop on Scanning Electron Microscopy of Human Reproduction was held in conjunction with the World Congress of Fertility and Sterility in Miami, Florida, April 1977. Some of the papers presented at this workshop appear in this volume with emphasis on problems of Andrology and Gynecology.

The editor wishes to express his thanks to all who contributed to the volume and to the various companies that supplied instrumentation and accessories for scanning electron microscopy. Thanks are also due to Ms. Jan Carter, Managing Editor of Ann Arbor Science Publishers, Inc. for providing excellent cooperation during the preparation of the series *Perspectives in Human Reproduction.*

Finally, I am also very grateful to Phillip Sherman for outstanding technical assistance and for the reproduction of the scanning electron micrographs.

E. S. E. Hafez
Detroit, Michigan

THIS VOLUME IS DEDICATED TO:

TAFIDA TAHER
and
SHARON NOONAN

CONTRIBUTORS

Allen, J. Department of Obstetrics-Gynecology, Birmingham Maternity Hospital, Queen Elizabeth Medical Centre, Edgbaston, Birmingham, B15-2TG, England

Barnhart, M. I. Department of Physiology, Wayne State University School of Medicine; and NIH Specialized Center of Thrombosis Research, Detroit, Michigan 48201

Connell, C. J. Department of Obstetrics-Gynecology, University of California, San Francisco, School of Medicine, San Francisco, California 94143

Elias, J. J. Department of Anatomy, University of California, San Francisco, California 94143

Fadel, H. E. Department of Obstetrics/Gynecology, School of Medicine, Medical College of Georgia, Augusta, Georgia 30902

Hafez, E. S. E. Reproductive Physiology Laboratories and Andrology Research Unit, C. S. Mott Center for Human Growth and Development, Wayne State University School of Medicine, Detroit, Michigan 48201

Jones, A. L. Cell Biology Section, Veterans Administration Hospital, Departments of Medicine & Anatomy, University of California, San Francisco, California 94122

Jordan, J. A. Department of Obstetrics-Gynecology, University of Birmingham, Edbaston, Birmingham, B15-2TG, England

Kamash, M. A. Reproductive Physiology Laboratories and Andrology Research Unit, C. S. Mott Center for Human Growth and Development, Wayne State University School of Medicine, Detroit, Michigan 48201

Lusher, J. M. Department of Pediatrics, and Children's Hospital of Michigan, Wayne State University School of Medicine, Detroit, Michigan 48201

Makabe, S. Department of Obstetrics-Gynecology, Toho University, Tokyo, Japan, Kanda 2nd Clinic, 20-14, 3-Chome, Nishiazabu, Minato-Ku, Tokyo, Japan

Noonan, S. M. Department of Pathology, Wayne State University School of Medicine, Detroit, Michigan 48201

Okamura, H. Department of Obstetrics-Gynecology, Kyoto University School of Medicine, Sakyoku, Kyoto, Japan

Oshima, M. Department of Obstetrics-Gynecology, Kyoto University School of Medicine, Sakyoku, Kyoto, Japan

Propping, D. Department of Obstetrics-Gynecology, University of Essen Medical School, Hufelandstrasse 55, 4300 Essen 1, West Germany

Sherman, P. S. Reproductive Physiology Laboratories and Andrology Research Unit, C. S. Mott Center for Human Growth and Development, Wayne State University School of Medicine, Detroit, Michigan 48201

Spring-Mills, E. Departments of Anatomy & Urology, State University of New York, Upstate Medical Center, 766 Irving Avenue, Syracuse, New York 13210

Tauber, P. Department of Obstetrics-Gynecology, University Hospital School of Medicine, 55 Hufelandstrasse, 43 Essen, West Germany

Watson, J. Edsel B. Ford Institute for Medical Research, Henry Ford Hospital, 2799 W. Grand Boulevard, Detroit, Michigan 48202

Weiss, L. Department of Pediatrics, Henry Ford Hospital, 2799 W. Grand Boulevard, Detroit, Michigan 48202

Zaneveld, L. J. D. Department of Physiology, College of Medicine, University of Illinois, P.O. Box 6998, 901 S. Wolcott St., Chicago, Illinois 60680

CONTENTS

SECTION I
METHODOLOGY

1. Introduction to Scanning Electron Microscopy (SEM) of Human Reproduction. 3
   E. S. E. Hafez and P. S. Sherman

2. Methodology of SEM 19
   E. S. E. Hafez and P. S. Sherman

SECTION II
ANDROLOGY

3. Spermatogenesis 39
   Carolyn J. Connell

4. Spermatozoa 57
   E. S. E. Hafez

5. The Seminal Coagulum. 69
   P. F. Tauber, D. Propping and L. J. D. Zaneveld

6. The Prostate Gland 89
   Elinor Spring-Mills and Albert L. Jones

SECTION III
GYNECOLOGY

7. The Ovary 99
   S. Makabe and E. S. E. Hafez

8. Oviduct-Uterus 107
   M. Oshima, H. Okamura and E. S. E. Hafez

9. Uterotubal Junction 133
   Hossam E. Fadel

10. The Cervix Uteri 147
    Josephine M. Allen and J. A. Jordan

11. Cervical Mucus 159
    L. J. D. Zaneveld, P. F. Tauber, D. Propping and
    E. S. E. Hafez

12. Amniotic Fluid Cells 177
    Sharon M. Noonan and Lester Weiss

SECTION IV
PRODUCTS OF CONCEPTION

13. The Mammary Glands 193
    Elinor Spring-Mills and Joel J. Elias

14. The Embryo and Fetus 203
    J. H. L. Watson and M. A. Kamash

15. Umbilical Cord & Neonatal Blood 217
    Marion I. Barnhart and Jeanne M. Lusher

INDEX 229

SECTION I

METHODOLOGY

CHAPTER 1

INTRODUCTION TO SCANNING ELECTRON MICROSCOPY (SEM) OF HUMAN REPRODUCTION

E. S. E. Hafez and P. S. Sherman

Extensive investigations have been carried out on the morphological, anatomical, ultrastructural and physiological aspects of the ovary (Hafez, 1978), oviduct (Hafez and Blandau, 1969; Hafez, 1974), uterus (Wynn, 1967), cervix (Blandau and Moghissi, 1973), vagina (Hafez and Evans, 1978) and placenta (Petry, 1963; Hoyes, 1970). Recently, the surface ultrastructures of the male and female reproductive systems, and products of conception have been studied by scanning electron microscopy (Hafez, 1975; Ludwig and Metzger, 1976). Several studies have been reported on the SEM structure of spermatozoa (Dott, 1969; Gould et al., 1971; Zaneveld et al., 1971; Lung and Bahr, 1972; Hafez and Kanagawa, 1973; Ludwig et al., 1972), human oviduct (Patek et al., 1972a,b, 1973), human endometrium (Ferenczy et al., 1972; Johannisson and Nilson, 1972) and human placenta and amnion (Ludwig, 1972; Ludwig et al., 1972). Scanning electron microscopy also has been used to study cervical and uterine carcinoma (Jordan and Williams, 1971; Ferenczy et al., 1972; Murphy et al., 1973; Granberg et al., 1975). Information gained from such studies has become an invaluable guide for the pathologist, physiologist, obstetrician, neonatologist and pediatrician.

Several methods have been used to preserve the soft biological tissues for SEM, including freeze-drying (Boyde and Barber, 1969; Small and Marszalik, 1969; Johanisson and Nilsson, 1972;

Patek *et al.*, 1972a), low-temperature evaporation (Arenberg *et al.*, 1970), critical point drying (Anderson, 1951; Horridge and Tamn, 1969; Ludwig, 1971; Ludwig and Metzger, 1971; Ludwig *et al.*, 1972), suspensions of spermatozoa on coverslip (Luse, 1970), and metal-plated coverslips taken from tissue culture (Cleveland and Schneider, 1969).

Critical point drying has been used successfully to study the scanning electron microscopic structure of a variety of tissues, *e.g.*, oviduct, uterus, cervix, vagina, epididymis, vas deferens, spermatozoa, cleaving eggs and placenta. Ciliary structures, such as the fimbriae and oviducts, are best used to evaluate the quality of the critical point drying technique. With scanning electron microscopy, one is able to observe large surface areas with high resolution and depth of field penetration. This provides valuable clinical information as to the topographic characteristics of the tissues.

TISSUE ORGANIZATION AND CELL TYPES

There are striking morphological differences between the tissue organization of the mucosa of different segments in the reproductive tract. In most organs, the cells were uniform in size, closely packed and resembled "cobblestone." In some instances, the surfaces of the cells were ill defined and covered with short microvilli and residual mucus (Figures 1.1 and 1.2).

Two basic cell types were observed in the epithelium of the reproductive tract: ciliated cells and noncilliated secretory cells. Ciliated cells were covered by cilia which overlapped the surface of secretory cells. The nonciliated cells had a dome-shaped surface and were covered with microvilli. The difference between the epithelium of various parts of the reproductive tract was particularly noted in the abundance of ciliated cells, and number of microvilli structures (Figures 1.3, 1.4 and 1.5).

There were remarkable morphological differences between surfaces of secretory and nonsecretory cells, and between normal and pathological cells. The surfaces of normal squamous cells in human cervices are arranged with interconnecting microridges, whereas the surface epithelium from carcinoma *in situ* is arranged in closely packed microvilli (Jordan and Williams, 1971).

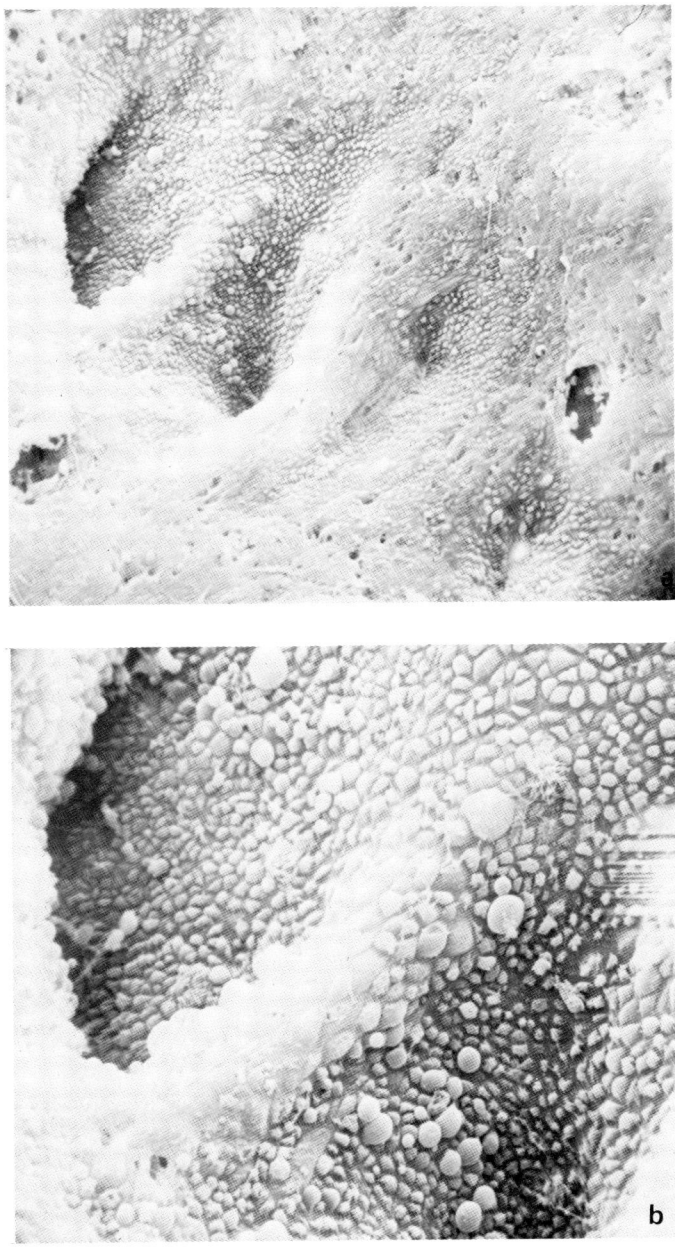

Figure 1.1. Opening of endometrial glands in human endometrium. Note large size of secretory cells at the mouth of the gland: (a) X240; (b) X575.

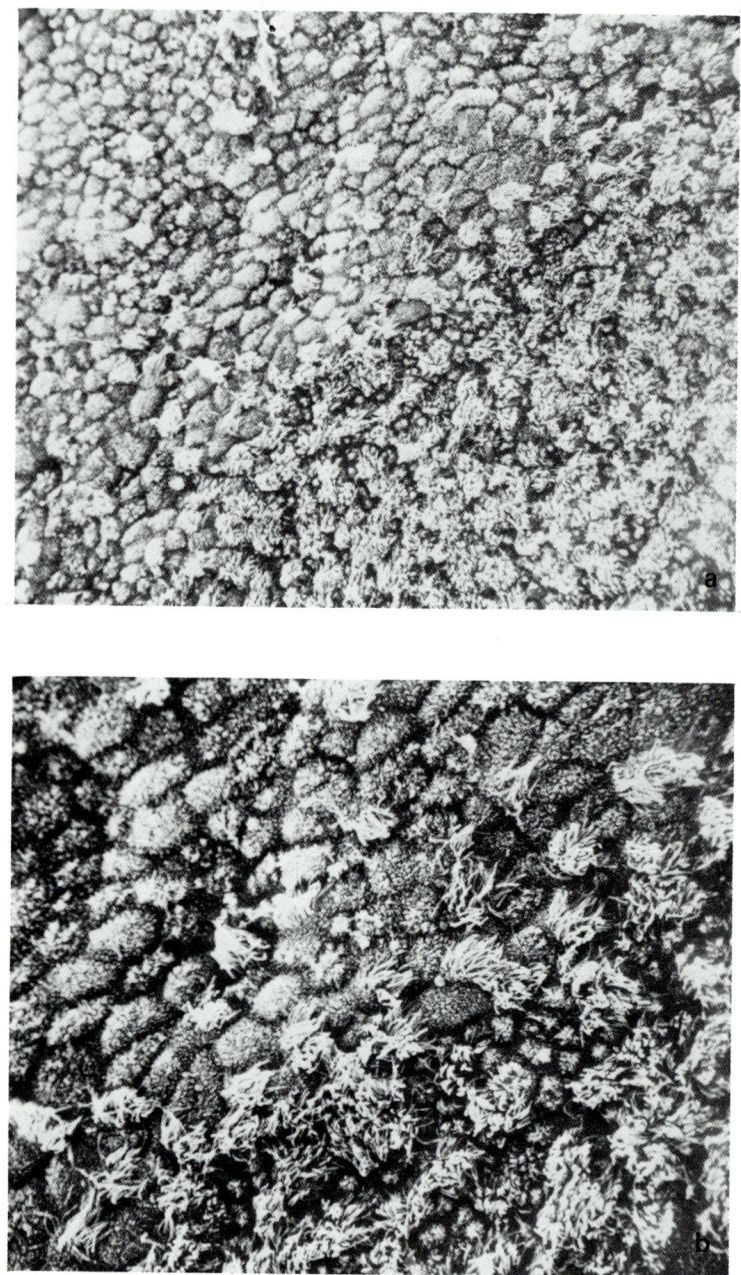

Figure 1.2. Human uterocervical junction. Uterine segment is at the top. Note higher frequency of ciliated cells in the endocervix at bottom: (a) X550; (b) X1100.

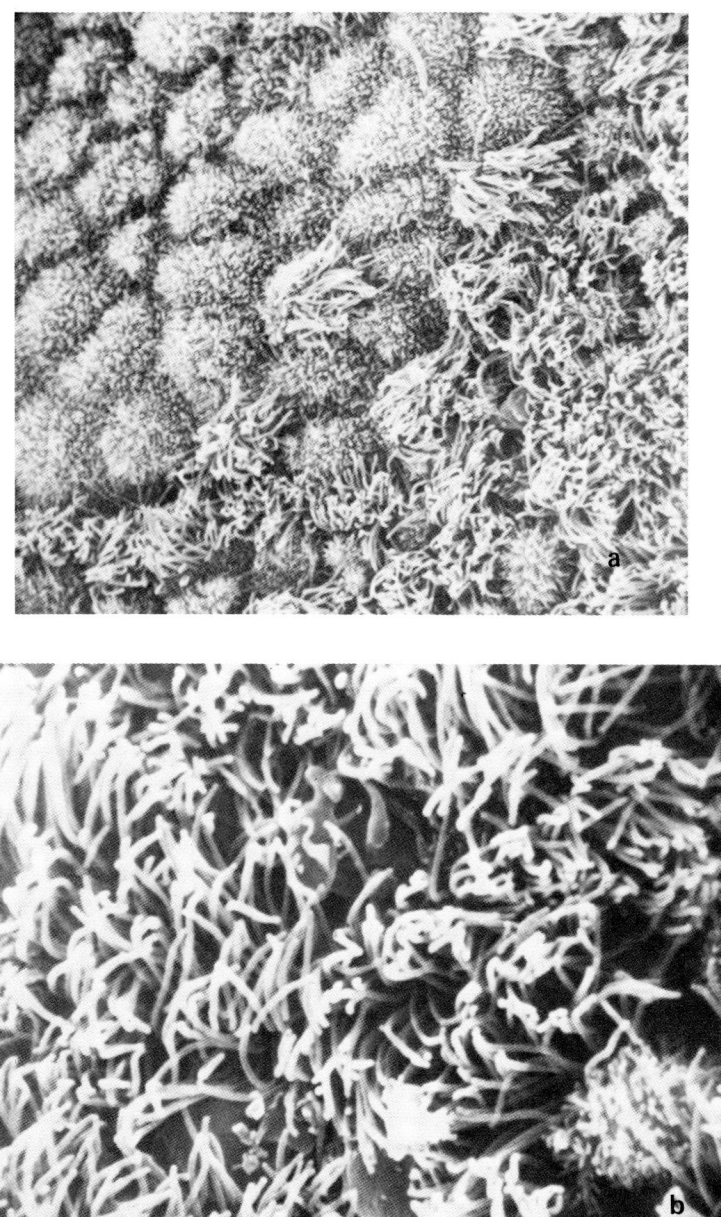

Figure 1.3. Ciliated cells in human uterocervical junction. Note abundance of ciliated cells and secretory cells with elongated microvilli: (a) X2200; (b) X5000.

Figure 1.4. Secretory and ciliated cells. Note morphological difference in microvilli of secretory cells (a) and kinocilia of ciliated cells (b, center cell). (a) X5000; (b) X5100.

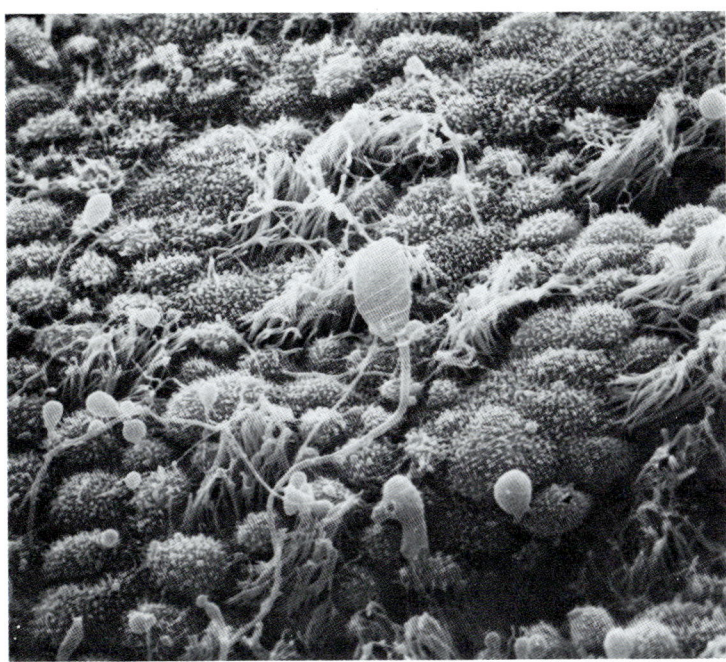

Figure 1.5. Sperm adhering to secretory and ciliated cells in the female reproductive tract. Note relative size of various cells and sperm head (X2000).

KINOCILIA

Kinocilia are motile and are found throughout the female reproductive tract except in the vagina (Figure 1.6). The directional beat of the cilia propels fluid currents toward the uterus. The basic cilia structure and its variations throughout the biological systems are well documented (Brenner, 1969; Fawcett, 1961; Satir, 1965).

Two types of ciliary motility have been recognized; an effective stroke and a recovery stroke (Hafez and Kanagawa, 1972). In the effective stroke, the cilia sweep rapidly in a stiff and slightly wavy motion. In the recovery stroke, the cilia bend near the basal body, and the degree of bending proceeds as a slow wave toward the tip. The recovery stroke carries the cilia back to the effective stroke position. Stroboscopic observation

Figure 1.6. Scanning electron micrographs of the vaginal epithelium. Note the rugae sloughing off of cells, distinct cell border, and microridges: (a) X480; (b) X5000. The SEM photograph was kindly provided by Dr. J. H. L. Watson of the Edsel B. Ford Institute for Medical

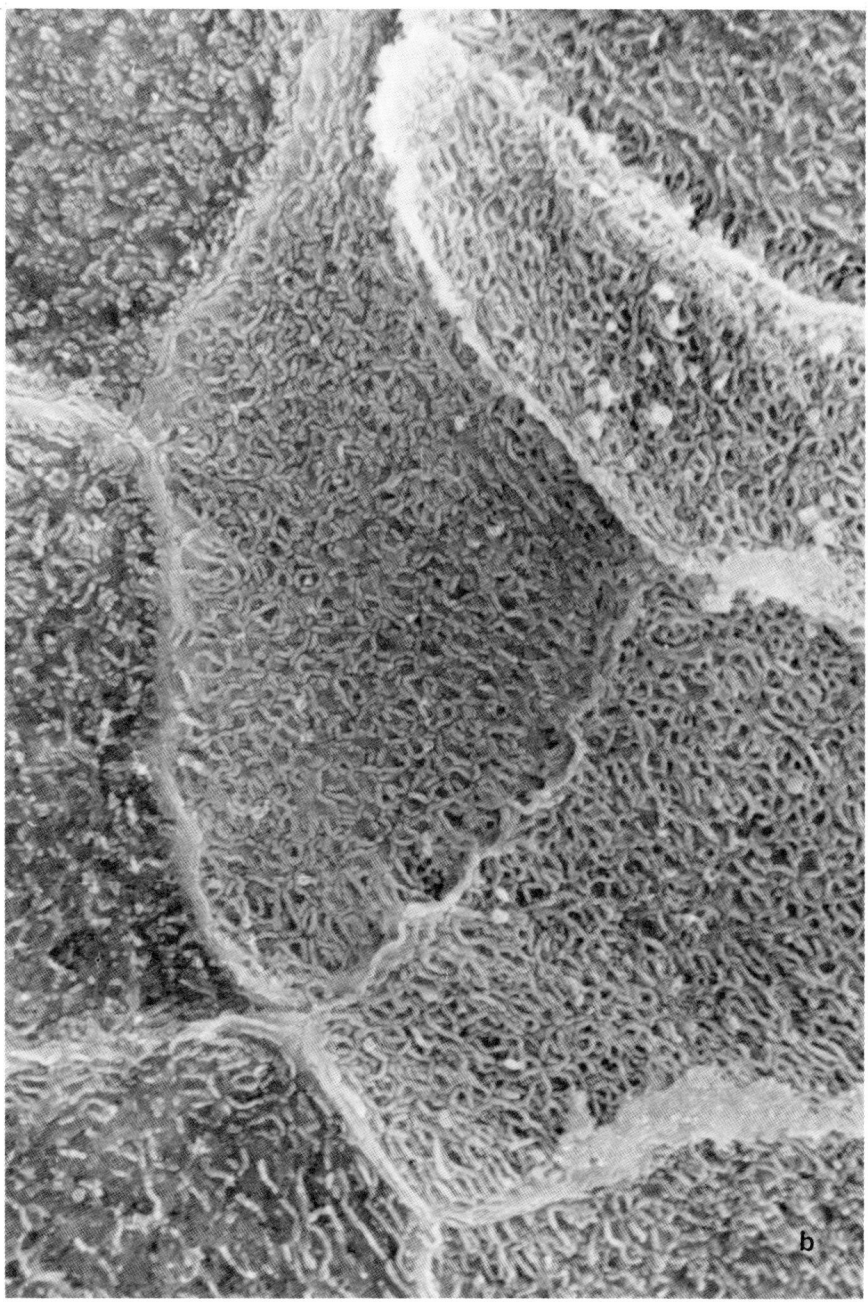

Research, Detroit, Michigan. Samples of human vagina were collected by Drs. M. Ismail and M. Gorrafa and processed by Dr. R. W. Steger and Mr. P. S. Sherman (Steger and Hafez, 1978).

of oviductal ciliated cells has shown that cilia beat approximately 1,200 times/minute (Borell et al., 1957). The kinocilia in the female reproductive tract beat rhythmically toward the vagina creating a directional flow of luminal fluids.

Ciliary activity is responsible for the movement of ova into the ostium of the fimbriated tip and through the upper ampulla. Concomitant with this, there is at the time of ovulation and during preliminary migration through the tube, a sharp increase in the intensity of muscular contractions (Koester, 1970; Aref and Hafez, 1973). Cilia of ampullary cells usually point toward the uterus. This is consistent with the evidence that ciliary currents in the ampulla sweep particles placed on the mucosal surfaces of rabbit and pig oviduct towards the ampullar-isthmic region (Gaddum-Rosse and Blandau, 1973). Although the ciliated cells in the ampulla appear to arise from crypt-like crevices because of the apical protrusion of the nonciliated cells, the ciliary tips have free access of movement into the lumen (Ludwig et al., 1972; Dirksen, 1975). The longitudinal mucosal folds with underlying musculature are characteristic of the ampullary isthmic junction. The folds are involved in the rotation of the egg and its subsequent transport (Blandau, 1969).

The cilia covering the oviductal epithelium play a major role in the release of secretory material from the adjacent secretory cells, and in the distribution of the secretions within the lumen. Infection of the oviduct is associated with the loss of ciliated cells in the oviduct (Reed and Boyde, 1972).

In the inflamed oviduct the ciliated cells are devoid of cilia, but retain a certain number of ciliary basal bodies, precursors of ciliary shafts. A decrease in cilia may lead to the accumulation of oviductal fluid and inflammatory exudate, which contributes to the agglutinations of tubal plicae and subsequent development of salpingitis (Ferenczy and Richart, 1974). It appears that the generation of cilia seems to be comparatively more conspicuous in fetal life.

CLINICAL APPLICATION

Scanning electron microscopy is a useful investigative tool which provides additional data for the study of the physiology and pathology of reproduction and hematological status of the newborn. SEM is also a valuable technique for diagnostic purposes, *e.g.*, infertility, carcinoma, fetal prematurity and certain birth defects (Figure 1.7). SEM characteristics of amniotic fluid cells (renal epithelium, sebaceous fat cells, pneumonocytes) may be of a clinical value in monitoring fetal maturity comparable to biochemical parameters.

SEM is also useful in the cytologic examination of spermatozoa for the diagnosis of male sterility. The use of ultrathin sections of spermatozoa for transmission electron microscopy gives only two-dimensional and fragmental information on the morphologic abnormalities of spermatozoa. Moreover, scanning electron microscopy is useful for the observation of free cells and subcellular bodies *in vivo* or *in vitro*.

SEM examination of various vaginal, cervical and endometrial lesions confirmed what has been previously described for normal, metaplastic, dysplastic and carcinoma *in situ* lesions of the vagina, ectocervix and uterus. A distinction was made between two types of squamous epithelium based on the presence of microridges rather than microvilli. Primary squamous cell carcinomas of the vagina, *i.e.*, cancers arising from mature, differentiated squamous epithelium, tend to retain their microvillous surface appearance. The SEM appearance of adenosis and adenocarcinoma of the vagina secondary to the *in-utero* exposure to Diethylstilbestrol was also illustrated.

SEM examination revealed the thrombogenetic pattern of placental infarction starting from pathologically altered syncytiotrophoblastic surface.

It may be useful to apply other techniques to verify the findings from scanning electron micrographs. Transmission electron microscopy, routine histological techniques, histochemical techniques and cinematography can be very helpful in certain studies. For example, to study tissues containing ciliated cells, the specimen could be divided into three portions. The first portion is immersed in tissue-cultured fluid and observed for cilia

14 Scanning Electron Microscopy of Human Reproduction

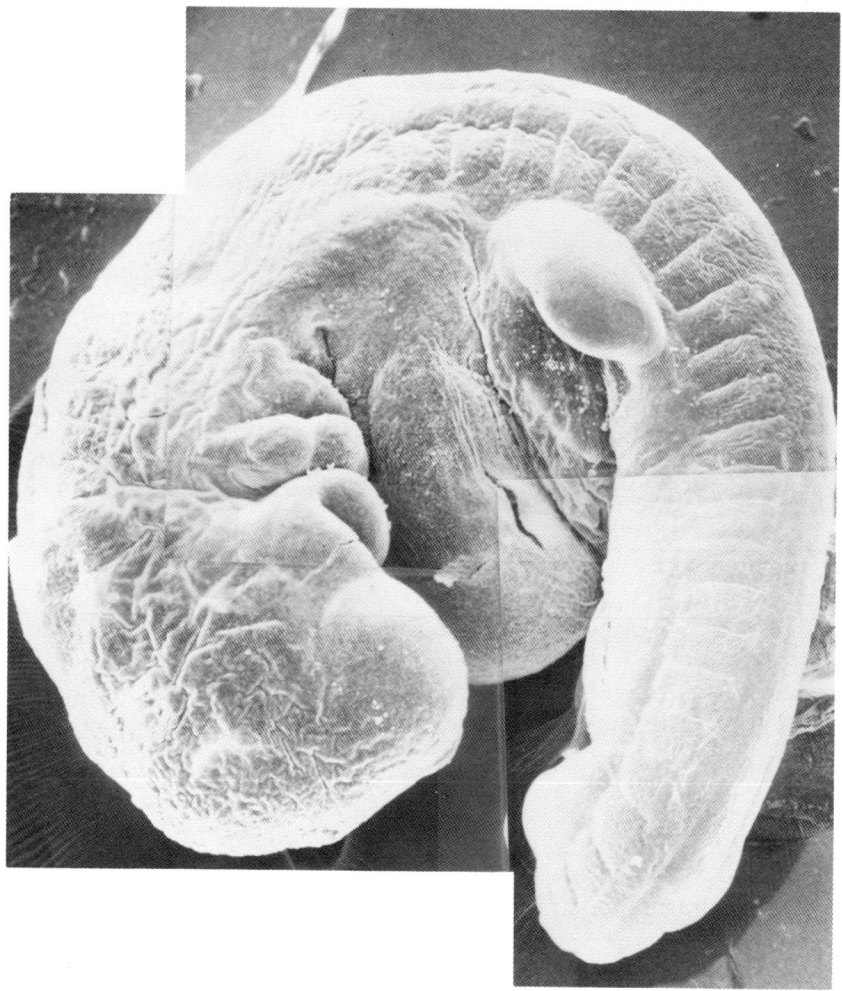

Figure 1.7. Human fetus, 28-day gestation, crown-rump 4.1 mm. Note the somites and lack of distinct cell boundary (X48).

beat, by Nomarski optics. A second portion, imbedded for histological sections, is stained with Masson triple stain to differentiate ciliated and secretory cells, thus making a differential cell count possible. The third portion is processed for scanning electron microscopy.

REFERENCES

Anderson, T. F. "Technique for the Preservation of Three-Dimensional Structure in Preparing Specimens for the Electron Microscope," *Trans. N.Y. Acad. Sci.* 13:130 (1951).

Aref, I. and E. S. E. Hafez. "Utero-oviductal Motility with Emphasis on Ova Transport," *Obstet. Gynecol. Survey* 28:679 (1973).

Arenberg, I., W. F. Marovitz and A. P. MacKenzie. "Preparative Techniques for the Study of Soft Biological Tissues in the Scanning Electron Microscope: A Comparison of Air Drying, Low-Temperature Evaporation, and Freeze-Drying," *Proc. 3rd Annual Cambridge Stereoscan Colloq.* 3:121 (1970).

Blandau, R. J. "Gamete Transport—Comparative Aspects," in *The Mammalian Oviduct* E. S. E. Hafez and R. J. Blandau, Eds. (Chicago: University of Chicago Press, 1969), p. 120.

Blandau, R. J. and K. S. Moghissi. *The Biology of the Cervix* (Chicago: University of Chicago Press, 1973).

Borell, U., O. Nilsson and A. Westman. "Ciliary Activity in the Rabbit Fallopian Tubes during Oestrus and after Copulation," *Acta. Obstet. Gynecol. Scand.* 36:22 (1957).

Boyde, A. and V. C. Barber. "Freeze-Drying Methods for the Scanning Electron Microscopical Study of the Protozoan *Spirostonnon ambigum* and the Statocyst of the Cephalopod Mollusc *Loligo Vulgaris*," *J. Cell Sci.* 4:223 (1969).

Brenner, R. M. "The Biology of Oviductal Cilia," in *The Mammalian Oviduct* E. S. E. Hafez and R. J. Blandau, Eds. (Chicago: University of Chicago Press, 1969), p. 203.

Cleveland, P. H. and C. W. Schneider. "A Simple Method of Preserving Occular Tissue for Scanning Electron Microscopy," *Vision Res.* 9:1401 (1969).

Dirksen, E. R. "The Oviduct (Mouse)," in *SEM Atlas of Mammalian Reproduction* E. S. E. Hafez, Ed. (Tokyo: Igaku Shoin, 1975).

Dott, H. M. "Preliminary Examination of Bull, Ram, and Rabbit Spermatozoa with the Steroscan Electron Microscope," *J. Reprod. Fertil.* 13:133 (1969).

Fawcett, D. W. "Cilia and Flagella," in *The Cell*, Vol. II, J. Brachet and A. E. Mirsky, Eds., (New York: Academic Press, 1961), p. 217.

Ferenczy, A, R. M. Richart, F. J. Agate, M. L. Pukerson and E. W. Dempsey. "Scanning Electron Microscopy of the Human Endometrial Surface Epithelium," *Fertil. Steril.* 23:515 (1972).

Ferenczy, A. and R. M. Richart. "Female Reproductive System," in *Dynamics of Scanning and Transmission Electron Microscopy* (New York: John Wiley & Sons, 1974), p. 213.

Gaddum-Rosse, P. and R. J. Blandau. "*In Vitro* Studies on Ciliary Activity within the Oviducts of the Rabbit and Pig," *Am. J. Anat.* 136:91 (1973).

Gould, K. G., L. J. D. Zaneveld and W. L. Williams. "Scanning Electron Microscopy of Mammalian Gametes," *Arch. Gynaekol.* 210:235 (1971).

Granberg, I., A. Ingelman-Sundberg, I. Joelsson, L. Nilsson and E. Patek. "Adenocarcinoma of the Uterine Corpus," in *SEM Atlas of Mammalian Reproduction* E. S. E. Hafez, Ed. (Tokyo: Igaku Shoin, 1975).

Hafez, E. S. E. and R. J. Blandau. *The Mammalian Oviduct. Comparative Biology and Methodology* (Chicago: University of Chicago Press, 1969).

Hafez, E. S. E. and T. N. Evans, Eds. *Human Vagina* (Amsterdam: Elsevier, 1978).

Hafez, E. S. E. and H. Kanagawa. "Ciliated Epithelium in the Uteri Cervix of the Macaque and Rabbit," *J. Reprod. Fertil.* 28:91 (1972).

Hafez, E. S. E. and H. Kanagawa. "Scanning Electron Microscopy of Human Monkey and Rabbit Spermatozoa," *Fertil. Steril.* 24 (1973).

Hafez, E. S. E., Ed. *Reproduction in Farm Animals*, 3rd ed. (Philadelphia: Lea & Febiger, 1974).

Hafez, E. S. E., Ed. *Scanning Electron Microscopy Atlas of Mammalian Reproduction* (Tokyo: Igaku Shoin, 1975).

Hafez, E. S. E., Ed. *Human Ovulation: Mechanisms, Prediction, Detection and Induction* (Amsterdam: Elsevier, 1978).

Horridge, G. A and S. L. Tamn. "Critical-Point Drying for Scanning Electron Microscopic Study of Ciliary Motion," *Science* 163:817 (1969).

Hoyes, A. D. "Ultrastructure of the Mesenchymal Layers of the Human Amnion in Early Pregnancy," *Am. Obstet. Gynecol.* 106:557 (1970).

Johanisson, E. and L. Nilsson. "Scanning Electron Microscopic Study of the Human Endometrium," *Fertil. Steril.* 23:613 (1972).

Jordan, J. A. and A. E. Williams. "Scanning Electron Microscopy in the Study of Cervical Neoplasia," *J. Obstet. Gynecol. Brit. Commonwlth.* 78:940 (1971).

Koester, H. "Ovum Transport," in *Mammalian Reproduction* H. Gibian and E. J. Plotz, Eds., (Berlin: Springer-Verlag, 1970), p. 189.

Ludwig, H. "Surface Structure of the Human Term Placenta and of the Uterine Wall Postpartum in the Screen Scan Electron Microscope," *Am. J. Obstet. Gynecol.* 111:328 (1971).

Ludwig, H. and H. Metzger. "Das Uterine Placentabett post partum in Rasterelektronenmikroskop, zugleich ein Beitrag zur Frage der extravasalen Fibrinbildung," *Arch. Gynaekol.* 210:251 (1971).

Ludwig, H. and H. Metzger. *The Human Female Reproductive Tract: A Scanning Electron Microscopic Atlas* (Berlin: Springer-Verlag, 1976).

Ludwig, H., H. Wolf and H. Metzger. "Zur Ultrastruktur der Tubeninnenflache im Rasterelektronenmikroskop," *Arch. Gynaekol.* 212:380 (1972).

Ludwig, H. "Rasterelektronenmikroskopische Befunde in Placenta, Einhauten und Plazentabett," in *Perinatale Medizin*, Vol. III, J. W. Dudenhausen and E. Saling, Eds. (1972), p. 237.

Lung, B. and G. F. Bahr. "Scanning Electron Microscopy of Critical-Point Dried Human Spermatozoa," *J. Reprod. Fertil.* 31:317 (1972).

Luse, S. A. "Preparation of Biologic Specimens for Scanning Electron Microscopy," *Proc. 3rd Annual Cambridge Stereoscan Colloq.* (1970), p. 149.

Murphy, J. F., J. M. Allen, J. A. Jordan and A. E. Williams. *6th Annual Scanning Electron Microscopy Symp.*, Chicago (1973), p. 605.

Patek, E., L. Nilsson and E. Johannisson. "Scanning Electron Microscopic Study of the Human Fallopian Tube. Report I. The Proliferative and Secretory Stages," *Fertil. Steril.* 23:459 (1972a).

Patek, E., L. Nilsson and E. Johannisson. "Fetal Life, Reproductive Life and Post-Menopause," *Fertil. Steril.* 23:719 (1972b).

Patek, E., L. Nilsson and E. Johannisson. "The Effect of Mid-Pregnancy and of Various Sterioids," *Fertil. Steril.* 24:819 (1973).

Petry, G. "Die Morphologie der ausserhalb der Plazenta bestehenden Feto-Maternen Kontakte," *Arch. Gynaekol.* 193:74 (1963).

Reed, S. E. and A. Boyde. "Organ Cultures of Respiratory Epithelium Infected with Rhinovirus or Parainfluenza Virus. Studies in Scanning Electron Microscopy," *Infect. Immun.* 6:68 (1972).

Satir, P. "Structure and Function in Cilia and Flagella: Protoplasmatologia," in *Handbuch der Protoplasmaforschung*, Vol. 3 (New York: Springer-Verlag, 1965).

Small, E. B. and D. S. Marszalik. "Scanning Electron Microscopy of Fixed, Frozen, and Dried Protozoa," *Science* 163:1064 (1969).

Steger, R. W. and E. S. E. Hafez. "Age-Related Changes in the Vagina," in *Human Vagina*, E. S. E. Hafez and T. N. Evans, Eds. (Amsterdam: Elsevier, 1978).

Wynn, R. M. *Cellular Biology of the Uterus* [New York: Appelton-Century-Crofts (Meredith Publishing Corp.), 1967].

Zaneveld, L. J. D., K. G. Gould, W. J. Humphreys and W. L. Williams. "Scanning Electron Microscopy of Mammalian Spermatozoa," *J. Reprod. Med.* 6:174 (1971).

CHAPTER 2

METHODOLOGY OF SEM

E. S. E. Hafez and P. S. Sherman

Several methods are used for pinning, fixation, dehydration, drying, mounting, metal coating and viewing with the scanning electron microscope (Hafez *et al.*, 1975; Ludwig and Metzger, 1976).

PINNING

Tissues (approximately 7 × 7 mm of epithelial surface) are pinned to a cork plate of 2-3 mm in thickness (or to Dow corning 780 sealant). Specimens attached to cork plates are floated tissue side down, in freshly prepared fixative.

After fixation, two or more segments of 3 × 3 mm are cut from the specimen. The sides to be examined are placed upward and pinned to a thin cork plate (1 mm thickness) using non-metallic pins (*e.g.*, plastic pins or insect spines). The cork plate is very thin so that during the procedures of dehydration and critical point drying only a small amount of the solution is absorbed into the cork. This thin cork plate should fit into the specimen baskets of the critical point drying apparatus (7.0 × 2.5 cm) (Figure 2.1).

Figure 2.1a. Accessories used for critical point drying (courtesy Ted Pella, Inc., P.O. Box 510, Tustin, California, 92680).

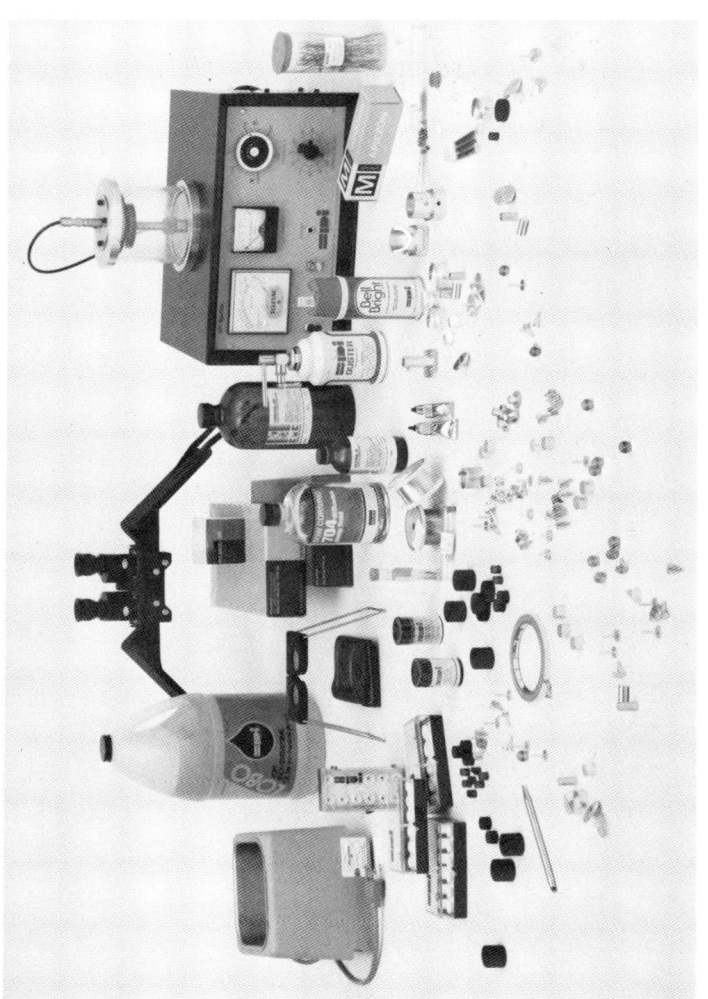

Figure 2.1b. Accessories used for processing biological samples for scanning electron microscopy. Note the Sputter evaporator on the top right (courtesy **SPI Supplies, P.O. Box 342, West Chester, PA 19380**).

FIXATION

A solution of 5% glutaraldehyde in 0.1 M phosphate buffer of pH 7.2 with an osmolarity of 699 mosm or 2.5% glutaraldehyde in the same buffer with osmolarity of 499 mosm give best results. A higher percentage of glutaraldehyde gives the best fixation, although it presents a greater risk of artifacts especially in very thin biopsies. A good indication of the optimal osmolarity is provided by the adjoining red corpuscles: if they maintain their bioconcave shape, the higher osmolarity has not adversely influenced them. Before immersing the tissues in the fixative, it is necessary to rinse them gently with an isotonic solution of Krebs-Ringer-glucose or physiological saline. The tissues are kept in the fixative for 24-48 hours. Tissue specimens covered with large quantities of mucus, such as the uterine cervix, are then transferred to a 0.2 M sucrose solution in 0.1 M phosphate buffer of pH 7.2 for a further 24 hours to dissolve and remove any mucus.

Tissues containing ciliated cells are best fixed in 2.5% cacodylate-buffered glutaraldehyde for 24 hours at 4°C. The time of fixation depends on the type of tissue.

DEHYDRATION AND DRYING

The tissues are dehydrated for 20-30 minutes in each of 30, 50, 70 and 90% alcohol; and for 10-20 minutes in each of 96% and absolute alcohol. Occasionally, the tissues are moved slightly. When critical point drying is used, the tissue is placed in amyl acetate for at least 30 minutes with occasional moving (Figure 2.2). If the specimen is prefixed in osmium tetroxide, the buffer in which the osmium is prepared has to be isotonic. This may then be followed by glutaraldehyde. The critical point drying method, first developed by Lampert and Koschorek (1971), involves several steps: (a) washing, (b) fixation, (c) dehydration, (d) substitution with liquid CO_2, (e) heating to supercritical temperatures, and (f) pressure release. Freon® (Cohen et al., 1968) and liquid CO_2 (Anderson, 1951) are most suitable for scanning electron microscopy, and no difference has been observed in the results by either of these methods.

Methodology 23

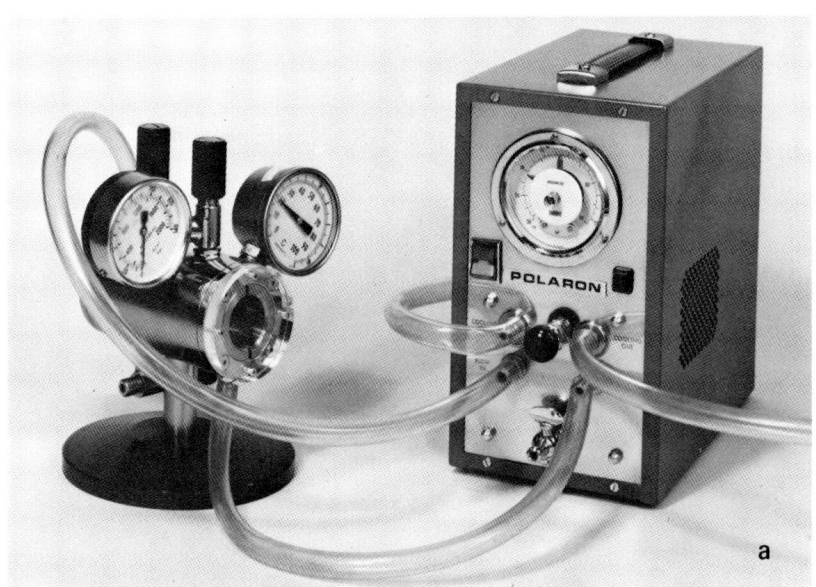

Figure 2.2 (a) Critical point dryer and (b) Sputter coater (courtesy Polaron Instrument Corp., 1202 Bethlehem Pike, Line Lexington, PA 18932).

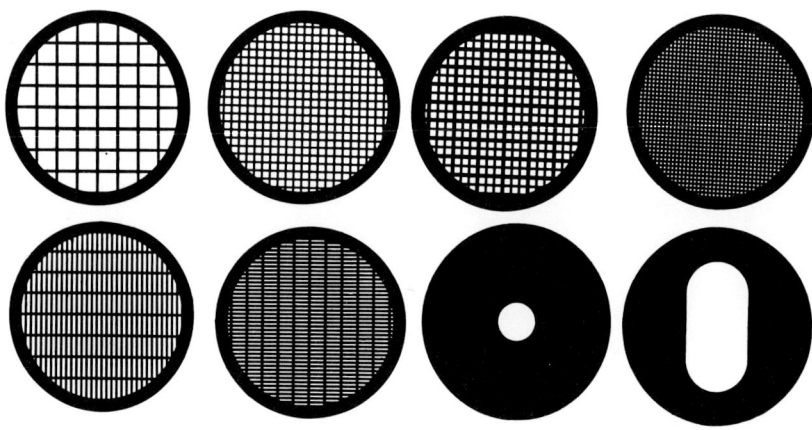
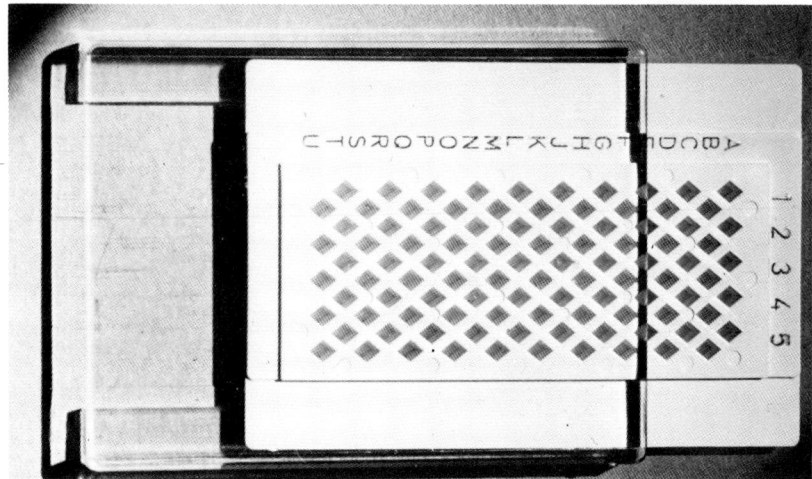

Figure 2.3. *Top.* Specimen Grids, for mounting sections for electron microscopy, are made of copper by an electrolytic process. Choice of two outer diameters for different types of electron microscopes and nine different mesh sizes. The four types having single openings are suitable for mounting serial sections.
Bottom. Specimen Grid Box protects delicate sections and supporting films and holds 100 grids in 20 rows of five diamond-shaped holes which take grids of O.D. 2.3 and 3 mm (courtesy LKB Instruments, 12221 Parklawn Drive, Rockville, Maryland 20852).

At the beginning of the drying process, the water cover is cooled using dry ice down to -15°C, to ensure that the carbon dioxide runs into the pressure chamber in liquid form. It is important that the cork plate containing the tissue be quickly transferred from the last bath of amyl acetate to the pressure chamber, which is immediately closed to avoid evaporation of amyl acetate. As the liquid CO_2 is added, the pressure increases in the chamber, and the color of the specimen changes from dark grey to white. The amyl acetate is removed by pressure release and slow refilling of the chamber with CO_2. The tissue is always covered with CO_2, thus the chamber is never completely emptied. The procedure, repeated 3-4 times, lasts about 45-60 minutes. The pressure chamber is cooled during the entire process of drying. When the tissue is free of amyl acetate, the water cover is gradually heated from 35-38°C. The pressure then rises and the CO_2 reaches the critical point where the liquid CO_2 is transformed into a gaseous state. The pressure is then slowly released until the pressure inside the chamber is equal to the pressure outside the chamber, at which time the specimen may be removed.

MOUNTING AND METAL COATING

Specimens are mounted on brass or aluminum slugs according to the type of scanning electron microscope to be used. A thin layer of silver glue is used, with an additional conducting rim of the same glue. The specimens are coated with carbon and gold in a vacuum evaporator at 5×10^{-5} torr. Electrical conductivity from the specimen surface to the holder is assured by painting the junction and sample edge with conductive silver paint.

Charging of nonconductive particles on or above the specimen surface of a nonconductive area of the specimen causes a disturbance of image formation, even if the specimen is sufficiently coated with a conducting layer. Sputtering of some of

26 *Scanning Electron Microscopy of Human Reproduction*

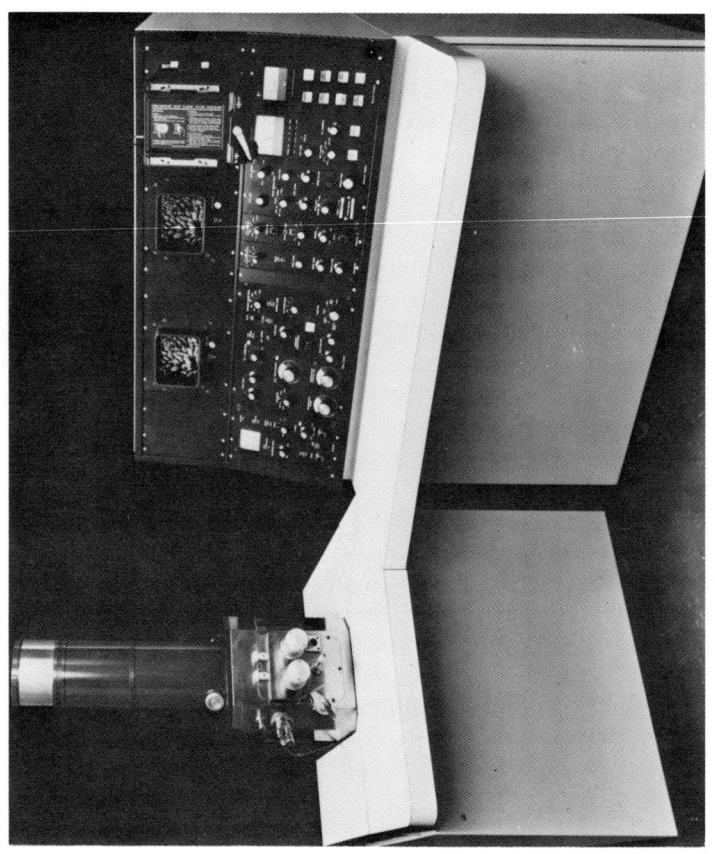

Figure 2.4. Stereoscan Scanning Electron Microscope S150 with continuous 1-40 kV operation. Note scanned gun alignment, totally ion pumped column and LaB_6 gun, digital focusing, alphanumerics, automatic focus and brightness and several electronic built-in features, available on many of the instruments.

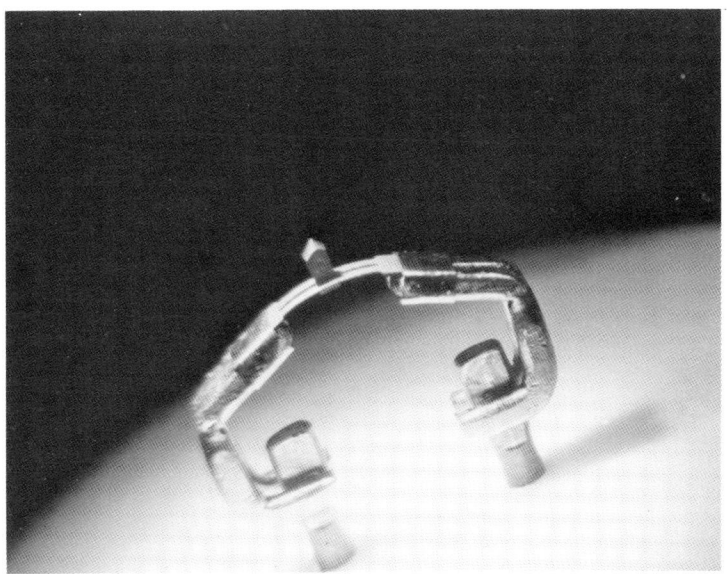

Figure 2.5. Plug-in replacement for tungsten filament for higher brightness, higher resolution, good signal/noise, small source size, low energy spread, easier driving and improved lifetime. It is an improved (single-crystal) version of the widely used LaB_6 cathode (polycrystalline) which is mounted on standard bases and fits most SEMs and probes. A cathode consists of a precision-machined grain-of-sand-sized Lanthanum Hexaboride block, mounted on a precision-etched 50-μm-thick arch-shaped carbon strip, which is in turn mounted in refractory metal holders. It is powered by I^2R heating using the original equipment power supply. No capital equipment purchases need be made, unless vacuum in the gun region is poorer than 10^{-6} torr. Typical operating parameters are about 2V and 2.5 A, or approximately 5 watts for high brightness operation; the drive impedance may be changed by use of different carbon strip dimensions. Actual power depends upon the size of the cathode block, the emission current needed and the desired life (Kimball Physics, Wilton, New Hampshire 03086).

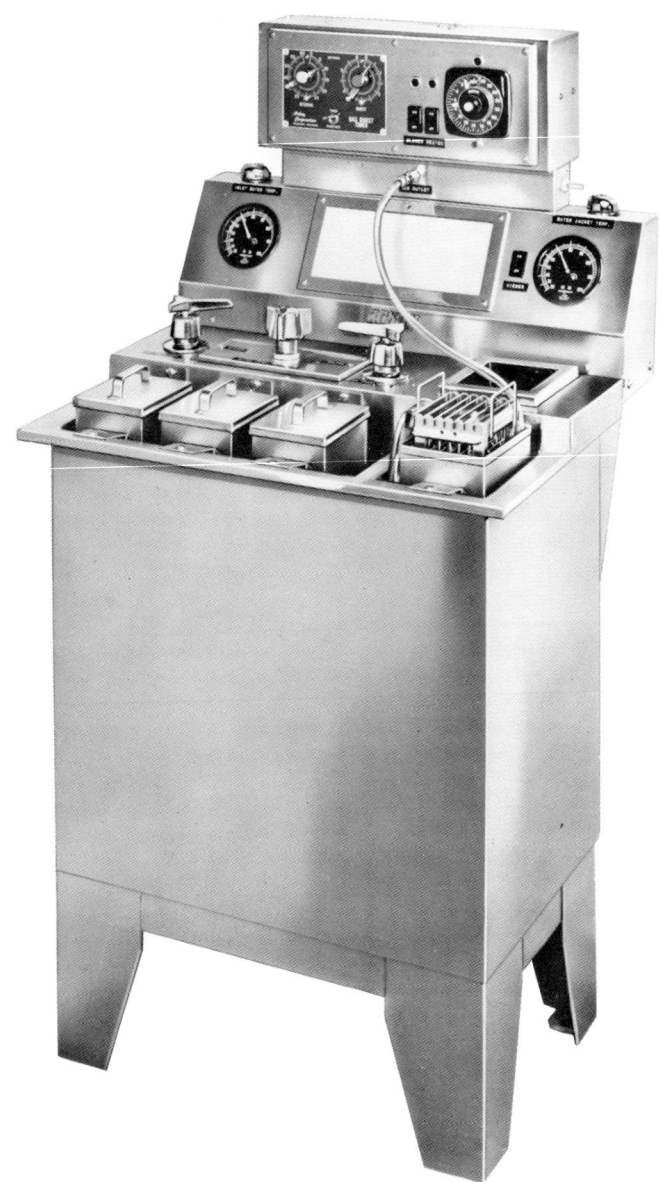

Figure 2.6. EM 410-4A photographic processor for polaroid SEM microphotographs. Note accessory water chiller which requires some circulating cold water from the laboratory (courtesy Arkay Co., 2285 Furst St., Milwaukee, Wisconsin 53204).

Figure 2.7. Stereoprojector with zoom (courtesy Ted Pella, Inc., P.O. Box 510, Tustin, California 92680).

the metal and compounds as carbides is more advantageous than the evaporation of these materials. The sputtering method, performed by an ion-gun is applied to shadow and produces thin films of metals (Figures 2.1 and 2.2). A practical solution to any possible problem of charging during viewing is to coat the specimen several times with metal until observation in the microscope microscopy is possible (Dirksen, 1975).

VIEWING WITH SEM

American, British and Japanese scanning electron microscopes are available. Some of them are supplied with electronic automatic features and special filaments (Figures 2.4 and 2.5). Several photographic processors are useful for polaroid negatives (Figure 2.6) and stereoprojectors are invaluable for projecting three-dimensional images during teaching and instruction (Figure 2.7). Other optional features to expand the basic functions of the scanning electron microscope are available on most instruments.

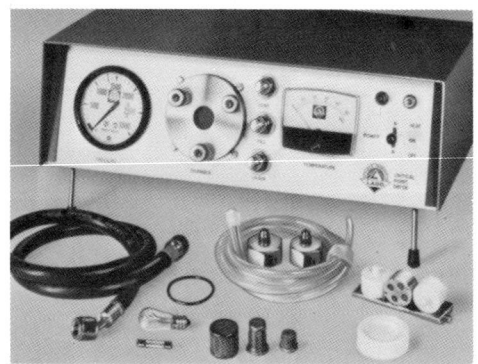
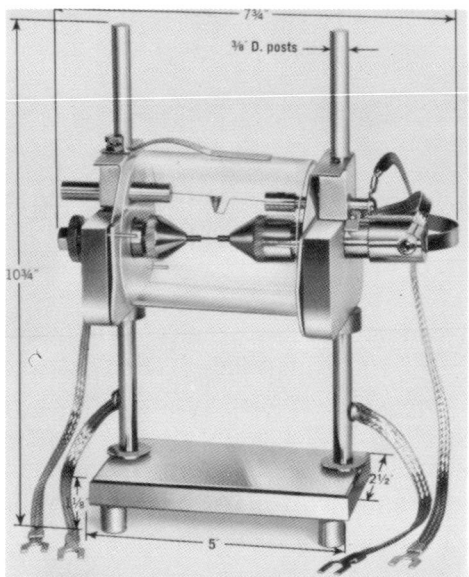
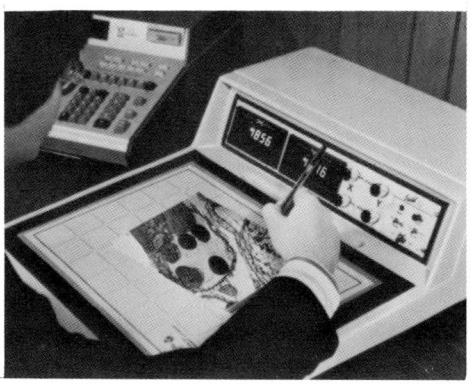

Methodology 31

Figure 2.8a. Critical point dryer, electrically heated, controlled and self-monitored. The chamber is designed to permit true bottom draining or independent upper venting. The large chamber allows processing of material ranging from the smallest samples up to and including six standard microscope slides. The internal side illumination of the chamber permits front-window viewing for monitoring of material being processed. All common critical point transitional fluids may be used with this dryer, including Freon and carbon dioxide. Accessories include specimen baskets, 8-sample holder, tank connecting hose, tank adapter fittings for different-sized compressed gas tanks, chamber gasket and Teflon® sealing tape.

Figure 2.8b. Shielded double-source evaporation unit made of heavy chrome-plated brass base with firmly positioned upright posts. Each post has an extension to hold: a carbon evaporation source, and a tungsten basket, metal evaporation source. Both sources are shielded by a single annealed Pyrex® tube with a carefully proportioned slot in the bottom through which the evaporated material flows to the specimen. The unit permits evaporation of the carbon and metal on one pump-down. The bell jar is kept clean, so accurate observation of the evaporation process is possible. Saves clean-up time.

Figure 2.8c. Graphic data digitizer which provides a simple user-oriented system for the analysis of all types of graphically presented images. This system is a useful tool in providing such parameters as areas, lengths, perimeters and diameters from almost any graphical source. It not only yields coordinate numbers usable directly for mathematical calculations, but can also transmit these data directly to calculators, computers or time-shared terminals, for instantaneous calculations. The unit may also be connected to any standard teletype for those who wish to store data on punched paper tape for processing in the future (courtesy Ladd Research Industries, P.O. Box 901, Burlington, Vermont 25401).

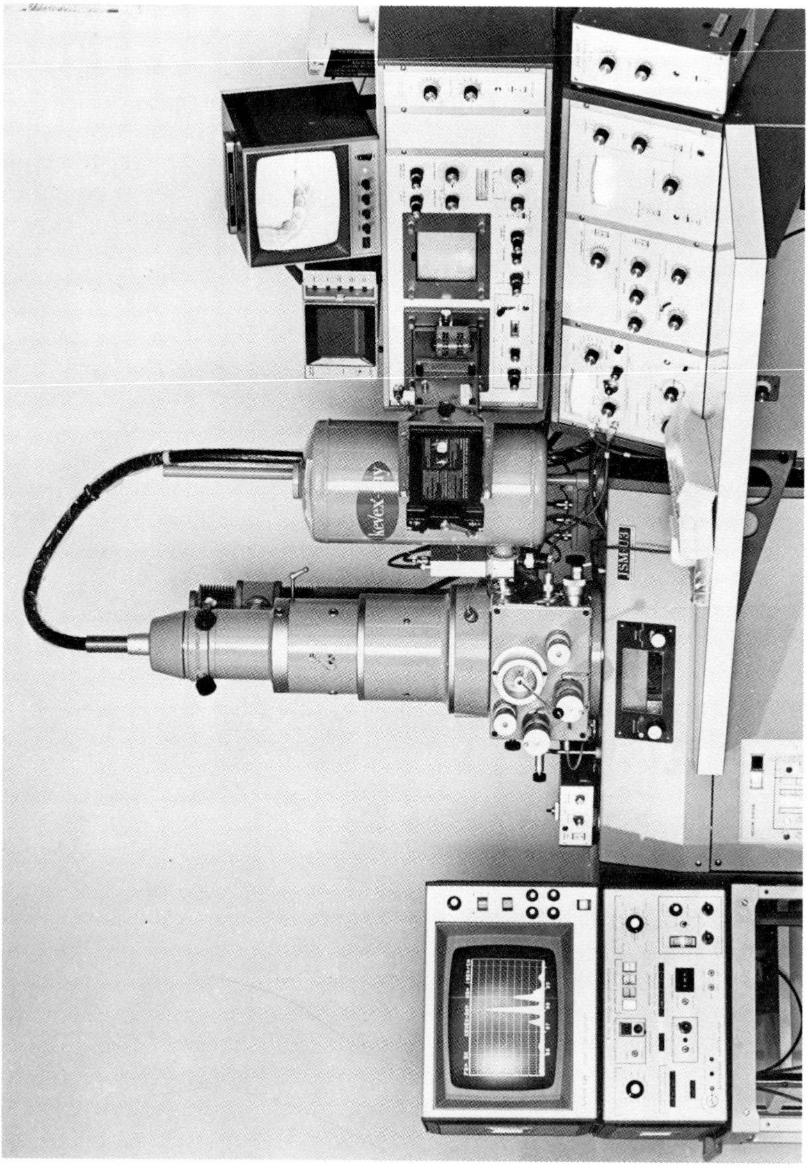

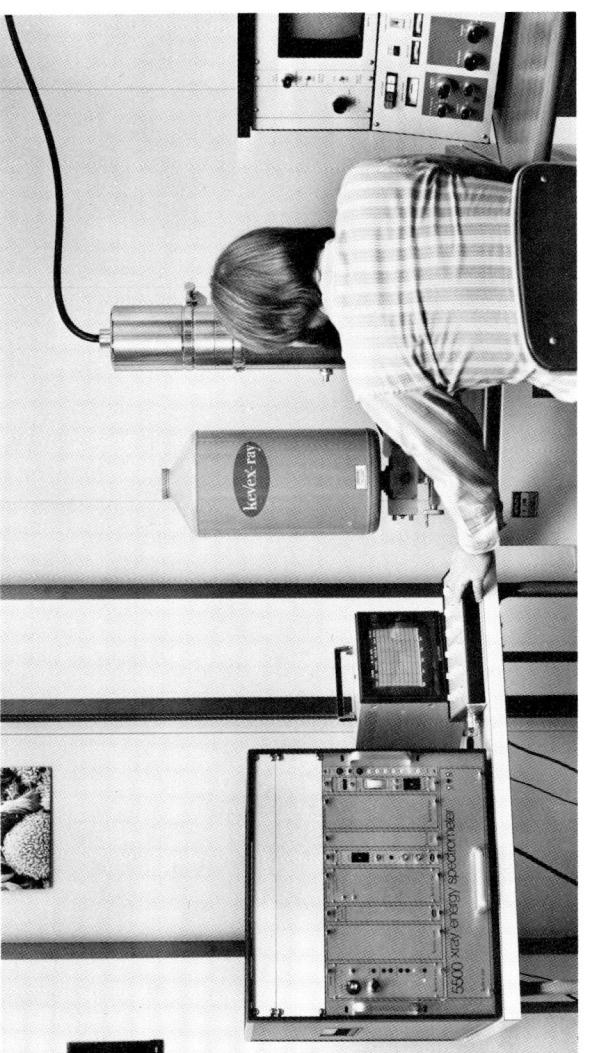

Figure 2.9. Micro-X analytical spectrometer provides various capabilities: XES (x-ray energy spectrometry); PHA (general-purpose pulse height analysis); and SEQ (sequential pulse counting, sampling of continuous analog waveforms, or averaging or repetitive transients). A simultaneous XES-SEQ mode allows an x-ray energy spectrum to be acquired in one memory group simultaneously with the acquisition of a sequential wavelength dispersive X-ray, Auger, electron energy loss, or other spectrum in a second memory group (courtesy Kevex, Analytical Instrument Division, Kevex Corporation, 1101 Chess Drive, P.O. Box 4050, Foster City, California 94404).

Figure 2.10. Computerized X-ray (Exac-1000) data analysis system for scanning electron microscopes and microprobes. The instrument utilizes pulse counting and silicon detectors (courtesy of Nuclear Equipment Corporation, 963A Industrial Road, San Carlos, California 94070).

On some SEM models a transmission mode may be obtained. In effect it is combining an SEM and a TEM into one instrument. The transmission option uses standard grids and supplies used in regular transmission (Figure 2.8). X-ray analysis for the multipurpose detection of elements in the sample specimen is also very useful (Figure 2.9). Computerized data analysis may be used with X-ray equipment (Figure 2.10).

REFERENCES

Anderson, T. F. "Technique for the Preservation of Three-Dimensional Structure in Preparing Specimens for the Electron Microscope," *Trans. N.Y. Acad. Sci.* 13:130 (1951).

Cohen, A. L., D. P. Marlow and G. E. Garner. "A Rapid Critical Point Method Using Fluorocarbons (Freons) as Intermediate Transitional Fluids," *J. Microscopie* 7:331 (1968).

Dirksen, E. R. "The Oviduct (Mouse)," in *SEM Atlas of Mammalian Reproduction* E. S. E. Hafez, Ed. (Tokyo: Igaku Shoin, 1975).
Hafez, E. S. E., M. I. Barnhart, H. Ludwig, J. Lusher, I. Joelsson, J. L. Daniel, A. I. Sherman, J. A. Jordan, H. Wolf, W. C. Stewart and F. C. Chrétien. "Scanning Electron Microscopy of Human Reproductive Physiology," in *Acta Obstet. Gynecol. Scand.*, Suppl. 40 (1975).
Ludwig, H. and H. Metzger. "The Human Female Reproductive Tract," in *A Scanning Electron Microscopic Atlas* (New York: Springer-Verlag, 1976).
Lampert, F. and F. Koschorek. "Elektronenmikroskopische Präparation biologischer Objekte ohne Dunnschnittechnik durch Oberflächenspreitung und kritische Punkt-Trocknung," *Z. Kinderheilk.* 111:29 (1971).

SECTION II

ANDROLOGY

CHAPTER 3

SPERMATOGENESIS

Carolyn J. Connell

Spermatogenesis, the production of testicular spermatozoa from spermatogonia, can be divided into four phases: (1) germ cell proliferation and renewal, (2) meiosis, (3) spermiogenesis, and (4) spermiation. The *proliferation* of germ cells first occurs during gestation but ceases during early postnatal life. Germ cell division does not reappear until ten or more years later when, concomitant with the prepubertal rise of testosterone, spermatogonia again divide mitotically to replace themselves and to give rise to the population of spermatocytes that will undergo meiosis. In *meiosis* the reductional and subsequent equational divisions give rise to spermatids that have half the chromosome number of spermatogonia and somatic cells. The round haploid cells, spermatids, undergo a complex series of biochemical and morphological changes resulting in the production of the elongate, flagellum-bearing, testicular spermatozoan. This process is referred to as *spermiogenesis*. At *spermiation*, the mature testicular spermatozoa are released from their enveloping Sertoli cells into the lumen of the seminiferous tubules. The nonmotile testicular spermatozoa, sperm, are swept from the testis into the excurrent canal system and epididymis for maturation into motile sperm. In this chapter the process of spermatogenesis will be discussed in relation to the structure and function of the Sertoli cells and the seminiferous tubules.

BASIC STRUCTURE OF THE SEMINIFEROUS TUBULES

Development of the Testis

In the early embryonic development of the gonads, gonocytes (primitive germ cells) migrate from the endoderm of the yolk sac to the genital ridge. Here they cease their migration and become surrounded usually by a single mesodermal cell (Witschi, 1951). The gonocytes and their enveloping cells form interconnected plate-like masses that are segregated from the interstitial tissue by a glycoprotein basement membrane. As testicular development advances in the fetus, the future seminiferous tubules become cord-like structures still surrounded by a basement membrane. Within the cords, immature Sertoli cells surround centrally located gonocytes (Gondos, 1977). These cells are actively mitotic. As maturation of the testis continues during gestation, the gonocytes move toward the periphery of the cords. After birth the gonocytes cease dividing mitotically and remain for a period of more than ten years in this undifferentiated state (Steinberger, 1974). With the onset of puberty and the concomitant increase of testosterone production, the gonocytes once again begin to divide mitotically and spermatogonia are formed. At about this time occluding junctions are established between the processes of Sertoli cells that overreach the basally located spermatogonia. These occluding junctions seal the adjacent Sertoli membranes together forming a permeability barrier known as the *blood-testis barrier*. Shortly after the establishment of the blood-testis barrier a lumen is formed in the center of the cords. The occluding junctions between Sertoli cells prevent the flow of interstitial fluid into the lumen as well as the flow of the luminal fluid out of the tubules. The fluid within the lumen of the seminiferous tubules differs in chemical composition from the surrounding interstitial and lymphatic fluid (Setchell and Waites, 1975). This special fluid environment is believed to be essential for the initiation of the first meiotic event (reduction division) and all future haploid development.

Structure of the Mature Testis

In the mature human testis the highly coiled and packed tubules do not form discrete loops as in the rodents, but the individual loops of tubules sometimes interconnect with each other as well as end in blind pouches (Johnson, 1934; Liang, 1966). The ends of the loops of seminiferous tubules connect with the straight tubules leading into the rete testis. Surrounding the tubules is a boundary tissue composed of peritubular cells coated with glycoproteins and overlaid by collagen fibers and a nonfibrillar matrix (Figure 3.1). The peritubular or myoid cells are smooth and muscle-like in appearance (Connell and Connell, 1977) and contractile in nature (Roosen-Runge, 1951; Clermont, 1958). These plate-like cells overlap each other to form a contractile sheath surrounding the seminiferous tubules (Figure 3.1). The contraction of the peritubular cells aids in the movement of the luminal contents into the excurrent canals. Sertoli cells form the stable nondividing epithelium of the testis and, with spermatogonia, constitute the only permanent cell population in the seminiferous tubules. Sertoli cells cease division before the initiation of meiosis (Steinberger and Steinberger, 1971a). Originally described by Sertoli in 1865, these unique cells are important in maintaining the structure of the tubules, in the metabolic exchange of materials from the interstitial tissue to the germinal elements (Mancini et al., 1963, 1965); in the coordination of spermatogenesis (Elftman, 1963); in phagocytosis (Roosen-Runge, 1955; Lacy, 1960; Carr et al., 1968; Black, 1971); in the production of proteins—androgen-binding protein (French and Ritzen, 1973; Fakunding et al., 1976), Müllerian inhibiting activity (Josso et al., 1975) and FSH-inhibitor-polypeptide (Franchimont et al., 1975); in the formation of the blood-testis barrier (Dym and Fawcett, 1970; Fawcett et al., 1970); in the movement of clones of haploid germinal cells from the basal component of the tubules to the lumen (Fawcett, 1975); and possibly in the elaboration of hormones—testosterone (Lacy et al., 1967; Van der Vusse et al., 1975) and estradiol (Dorrington and Armstrong, 1975). These conically shaped cells (Figure 3.2) sit on the basement membrane and are the only cells that extend from the basement membrane to the lumen. They have many

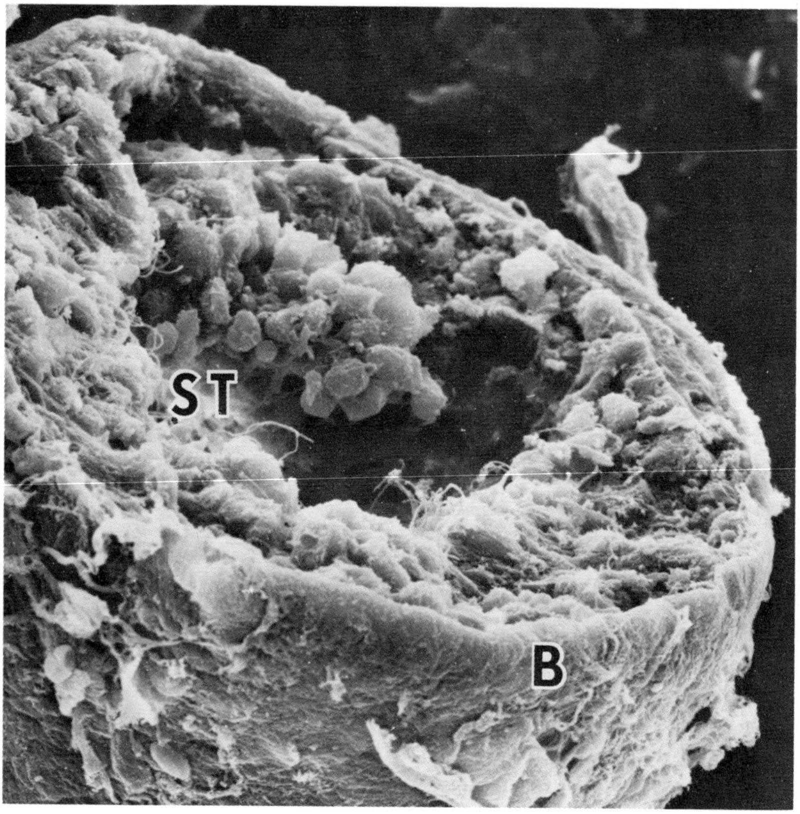

Figure 3.1. This SEM shows a cross-sectional view of a seminiferous tubule (ST) from a testicular biopsy of a 40-year-old man. Several "stages" of spermatogenesis are represented in this cross section. The seminiferous tubule is encased in a muscular boundary tissue (B) composed of myoid cells, collagen fibers and a glycoprotein matrix (X730).

highly interdigitating lateral processes. The germ cells are enclosed within thin sheet-like extensions of the cytoplasm of Sertoli cells. The appearance of human Sertoli cells in thin sections is consistent with a highly synthetically active cell (Nagano, 1966; Fawcett, 1975; Connell, 1978) and is similar to Sertoli cells of other mammals. The Sertoli cell plays a central role in the formation of the blood-testis barrier, spermiogenesis, and spermiation. A single Sertoli cell can be associated with three or more generations of germ cells and must therefore have a

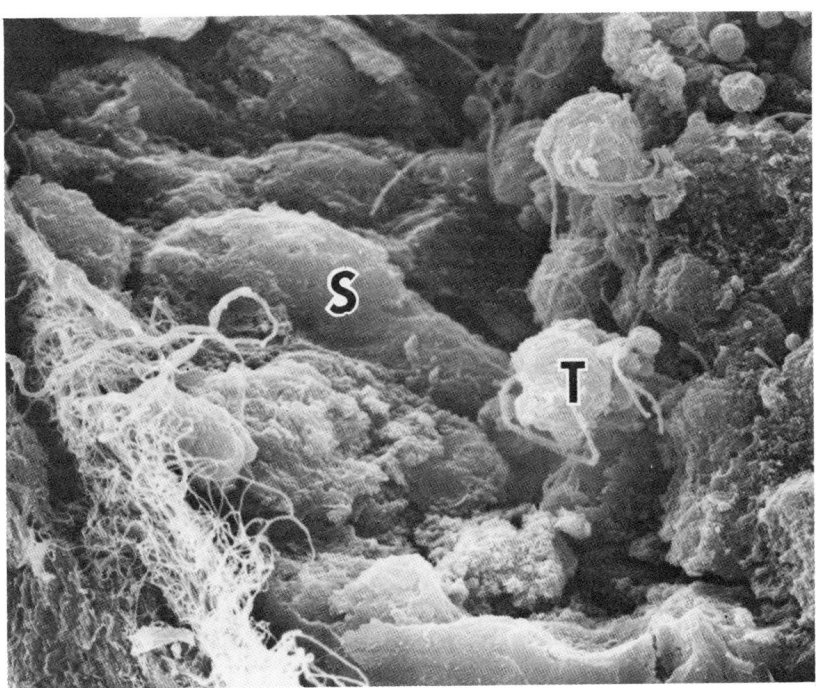

Figure 3.2. The conical Sertoli cells (S) extend from the basement membrane, here obscured by collagen fibers, to the lumen of the tubule. Round spermatids with flagellae are shown at the right (T). This tissue is from a testicular biopsy of an adult vasectomized man (X1550).

significant regional differentiation of its cytoplasm. The spermatogonia, the diploid stem cells from which all sperm are eventually derived, rest on the basement membrane. The diploid spermatocytes are located directly above the spermatogonia. The haploid germ cells, spermatids, are situated luminal to the diploid cells and are separated from them by bands of Sertoli cell cytoplasm that are sealed together by occluding junctions (Fawcett, 1975). In the tubules there are specific cell associations or *stages* of germ cells (Clermont, 1963). In any given cross section of human seminiferous tubules several (usually three or four) stages are present. This is in contrast to the condition observed in most mammalian species where a single stage of germ cell associations is present in a cross section and the stages of development proceed in an orderly arrangement from one end of a tubule to the other (Steinberger and Steinberger, 1975). This orderly arrangement has been referred to as the "cycle of the seminiferous epithelium" (Leblond and Clermont, 1952; Roosen-Runge, 1952). In rodents the tubules have only two connections: each tubule end is connected with the rete testis (Clermont and Huckins, 1961), a structure connected to the epididymis by way of the ductuli efferentes. Analysis of cell composition of the seminiferous epithelium from the connection of the seminiferous tubules to the rete testis reveals that a series of specific cell associations occur in consecutive stages along the tubule so that adjacent segments are either "less or more advanced by a single stage and the development always proceeds in one direction" (Perey et al., 1961). The tubules, when examined in a direction distal to the connection with the rete testis, show progressively less advanced stages of spermatogenesis. This situation reverses direction at the middle of the loop of the tubule and a descending order of maturation is noted as one progresses back to the rete testis (Steinberger and Steinberger, 1975). The wave concept in human seminiferous tubules is difficult to demonstrate because of the highly irregular cellular composition and topographic distribution (Clermont, 1963; Steinberger and Tjioe, 1968). In man, morphological and autoradiographic studies suggest that a small group of spermatogonia form a clone which develops independently of surrounding groups of cells (Chowdhury, 1971) rather than in waves as in rodents. In man, the time required for a spermatogonium to become a testicular spermatozoan has been reported as 64 days (Heller and Clermont, 1963).

GENERAL DESCRIPTION OF SPERMATOGENESIS

Spermatogonial Proliferation and Renewal

In the human testis the transformation of gonocytes into spermatogonia does not occur until puberty when androgen production significantly increases (Steinberger, 1974). Thus, the gonocytes remain in *mitotic arrest* for more than ten years and do not enter the process of transformation until sufficient testosterone appears to stimulate the process of differentiation.

The adult form of immature germ cells, the spermatogonia, occupy a position at the periphery of the seminiferous tubules adjacent to the basement membrane. Some of these spermatogonia differentiate and enter the pool of spermatogonia that are needed to replace the maturing germ cells in the continuously proliferating germinal epithelium (Courot *et al.*, 1970) while others are retained as *stem* cells. The mechanism by which the stem cells renew themselves and yet produce other spermatogonia that enter the process of spermatogenesis is still unclear, although several concepts have been developed to describe this process (Clermont, 1972; Steinberger and Steinberger, 1972; Oakberg and Huckins, 1976). The spermatogenic process is in part under the control of pituitary hormones and androgens (Steinberger and Steinberger, 1971b). The mitotic divisions of the differentiating spermatogonia do not result in the production of a mass of separate cells but rather a cluster or *clone* of cells connected one to another by bridges of cytoplasm (Figure 3.3) representing the progeny of a single spermatogonium. The intercellular bridges result from incomplete cytokinesis, cell separation, after division rather than cell fusion (Moens and Go, 1972). Cytoplasmic bridges of this type have been reported only in germ cell lines and the synchronous development of germ cells in a clone is ascribed to this cytoplasmic continuity. The cytoplasm bridges are retained throughout the development of the clone. Only at spermiation do the germ cells become separated. After the spermatogonia have entered into the process of differentiation, further morphological as well as biochemical changes occur (Bellvé *et al.*, 1977). The diploid spermatogonia enlarge in size and with the other members of the clone enter the reduction division or meiosis.

Meiosis

Human spermatogonia, like somatic cells, contain two sets of 23 chromosomes, or 46 chromosomes. Mature sperm as well as mature eggs contain but a single set of 23 chromosomes so

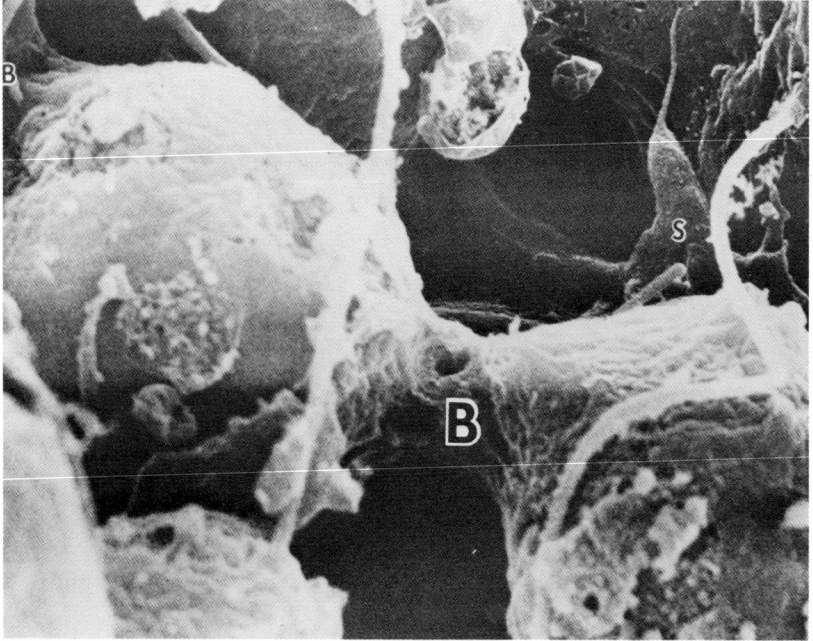

Figure 3.3. Clones of haploid germinal cells are connected by intercellular bridges (B) that are maintained during the entire process of spermiogenesis. This specimen is of canine origin (X9500).

that at fertilization the fusion of the male and female pronuclei does not result in the doubling of the chromosome number. There is insufficient space here to detail the complex process of meiosis and the reader is referred to a standard histology or genetics text for that description. In brief, however, the diploid spermatogonia enlarge in size and with the other members of the clone develop into primary spermatocytes while still within the basal region of the tubules. After moving from the basal to the adluminal compartment, the diploid spermatocytes undergo a precise cycle of nuclear events which culminate in the first meiotic division—the reduction division. The succeeding equational division produces two secondary spermatocytes from each primary spermatocyte. Secondary spermatocytes are smaller in size than primary spermatocytes and rapidly divide again to form round spermatids.

Spermiogenesis

Golgi vesicles containing a number of lytic enzymes coalesce and condense over the spermatid nucleus to form an acrosomal cap (Zaneveld, 1975). The progressively condensing chromatin of the spermatid nucleus is accompanied by an elongation of the nucleus and its overlying acrosome and the concomitant development of the spermatid flagellum. The appearance of the acrosomal cap is commonly used as a marker for the stage of development of the spermatid. In thin sections of round spermatids the presence of the sperm tail or flagellum is often overlooked because the tail is not in the same plane of section as the acrosomal cap. However, sperm tails are obvious at the round spermatid stage in scanning electron micrographs (Figure 3.4).

Spermiation

As elongation of the spermatid continues, nearly all the nuclear portion of the spermatid is freed of its Sertoli cell investments; however, an arm of spermatid cytoplasm extends from the neck region (Figures 3.5 and 3.6) deep into the Sertoli cell cytoplasm acting as an anchor or holdfast device. In addition to this holdfast attachment there are other types of attachments between spermatids and Sertoli cells. There are septate junctions between the acrosomal region of the spermatid and its Sertoli cell as well as bulbous extensions (Figure 3.8) of plasma membranes from the acrosome region into the Sertoli cell cytoplasm (Connell, unpublished observations; Russell and Clermont, 1976; Connell, 1975). The arm of spermatid cytoplasm that extends deep into the Sertoli cell cytoplasm breaks near the surface of the encompassing Sertoli cell and the portion retained within the Sertoli cell is known as the residual cytoplasm. The residual cytoplasm undergoes digestion and is phagocytized by the Sertoli cell. Once incorporated into the Sertoli cell cytoplasm, the spherical bodies containing the remnants of spermatid cytoplasm are known as residual bodies. The spermatid cytoplasm retained at the neck region is referred to as the cytoplasmic or protoplasmic droplet. The cytoplasmic droplet usually disappears during sperm maturation (Phillips, 1974).

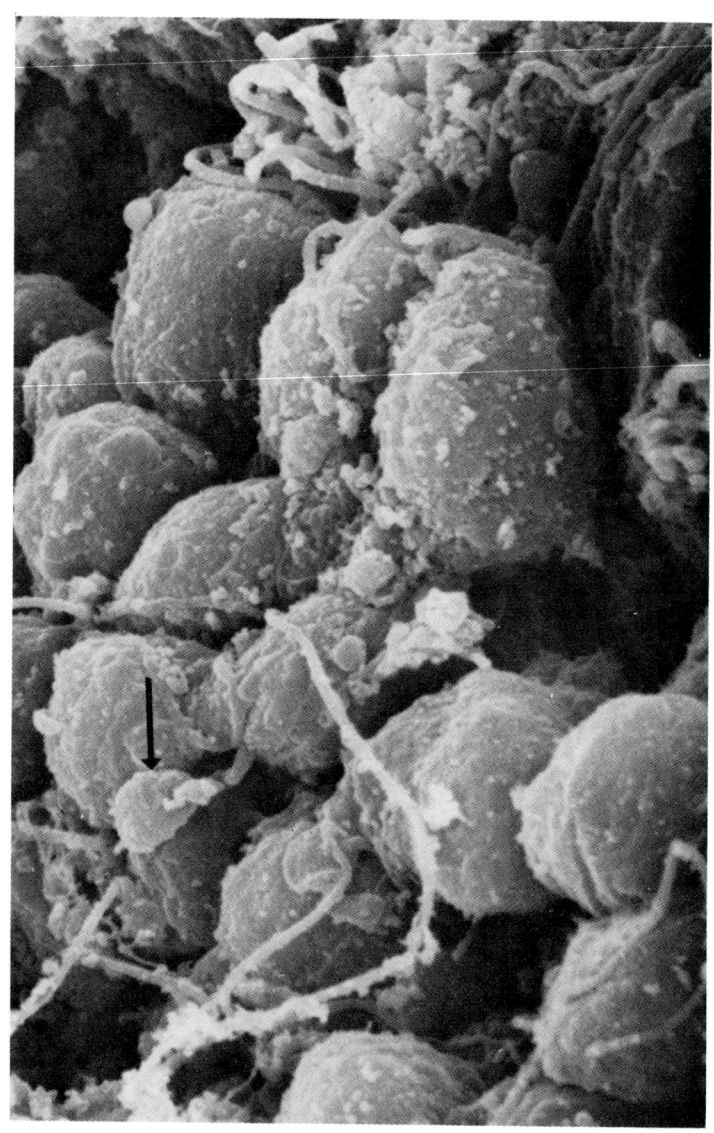

Figure 3.4. This SEM of human testis shows a luminal view of a clone of round spermatids. The young spermatids are adjacent to a region of mature testicular spermatozoa. One spermatozoan has been detached and artifactually attached to a spermatid (→) (X4500).

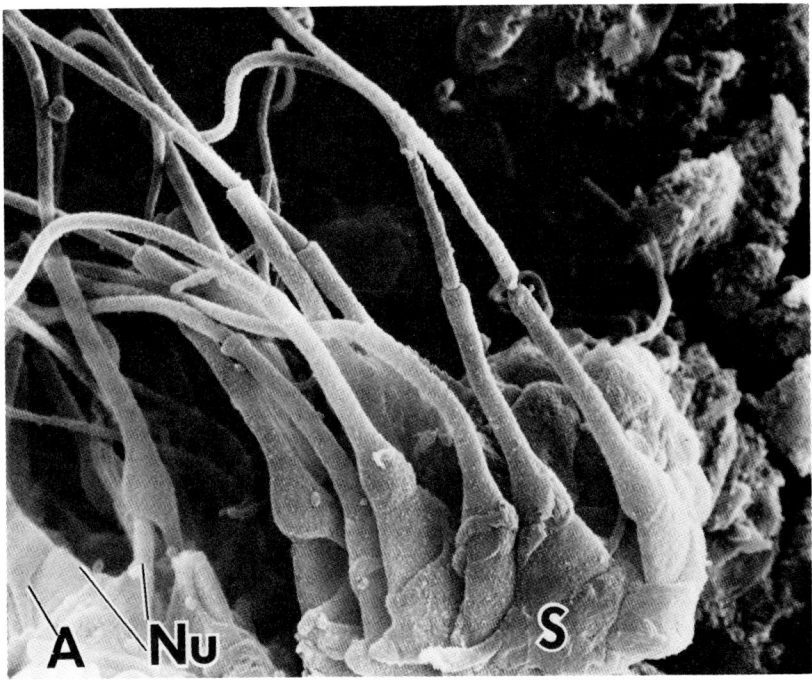

Figure 3.5. This SEM shows a clone of mid-late elongate spermatids from canine testis. Note their insertion into the Sertoli (S) cytoplasm. The spermatids on the left side of the micrograph clearly show the relationship of the nucleus (Nu) and the cytoplasm attachment (A) of the spermatid to ther Sertoli cell (X4400).

THE ROLE OF THE SERTOLI CELL IN SPERMIOGENESIS

Sertoli cells undergo alterations in shape, location of nucleus and cytoplasmic contact with the germ cell elements during spermatogenesis. The developing spermatocytes are attached to the Sertoli cells by small desmosomes (Connell, unpublished observations) and are moved, passively, from their basal location by the Sertoli cells. Sheets of Sertoli cell cytoplasm are interposed between the differentiating spermatocytes and the spermatogonia. A complex junction forms between adjacent Sertoli cells (Connell, 1978; Fawcett, 1975). Within this complex junction

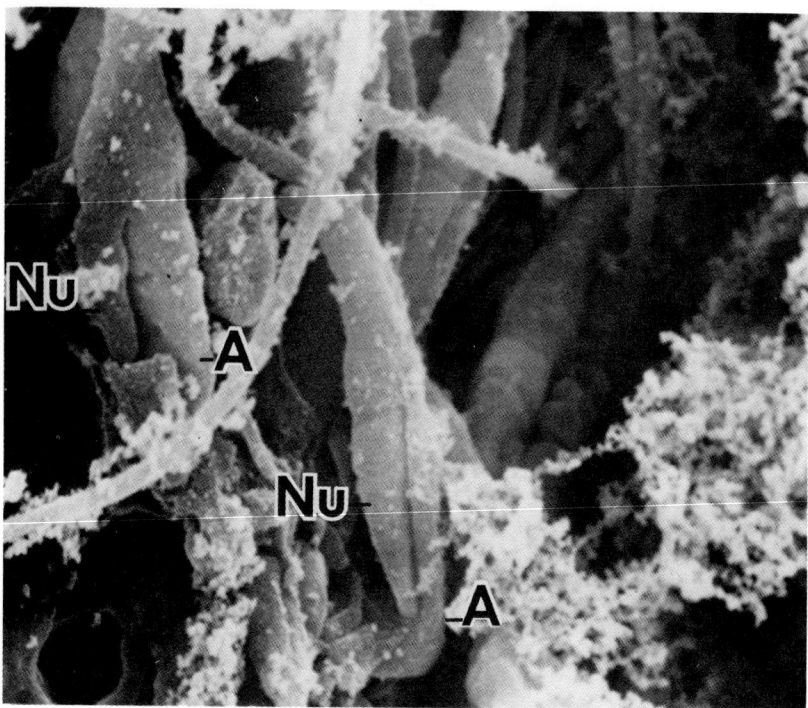

Figure 3.6. This SEM shows in human testis spermatid nuclei (Nu) and cytoplasm attachments (A) in late elongate stage spermatids. Only the cytoplasmic attachment appears to be holding the spermatids to the Sertoli cell cytoplasm (X8200).

are occluding junctions that effectively form a seal to the flow of materials from the interstitial fluid to the lumen of the seminiferous tubules. These occluding junctions, located above the spermatogonia and preleptolene spermatocytes, partition the seminiferous tubules into a basal and an adluminal compartment (Fawcett, 1975). This compartmentalization of the tubules results in three populations of cells: (a) those exposed only to the interstitial fluid, spermatogonia and the preleptolene spermatocytes; (b) those exposed only to the luminal fluid, postleptolene spermatocytes and spermatids; and (c) those exposed to both fluid environments, Sertoli cells (Figure 3.7). Sertoli cells extend from the basement membrane to the lumen and enfold the developing germ cells within their cytoplasm. Any hormonal

Andrology 51

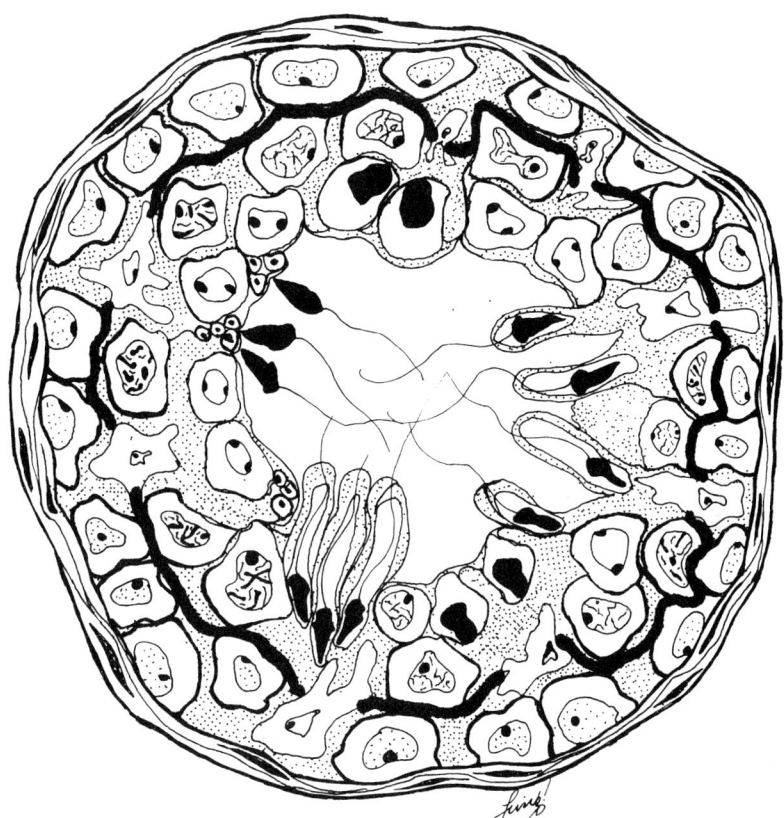

Figure 3.7. This diagram shows an idealized cross section of a human seminiferous tubule and depicts several stages of germ cell associations. A dense line indicates the approximate location of the blood-testis barrier. This barrier divides the tubules into a basal and an adluminal compartment. Only the Sertoli cells extend from the base of the tubules to the lumen. The stippled areas represent Sertoli cell cytoplasm. The tubule is surrounded by a muscular boundary tissue.

or nutritive change in the interstitial fluid will be modulated by the movement of these substances through the Sertoli cell cytoplasm before passing into the luminal compartment and affecting the haploid cells. This barrier protects the rest of the body from the effect of the unique spermatid antigens (Millette and Bellvé, 1977).

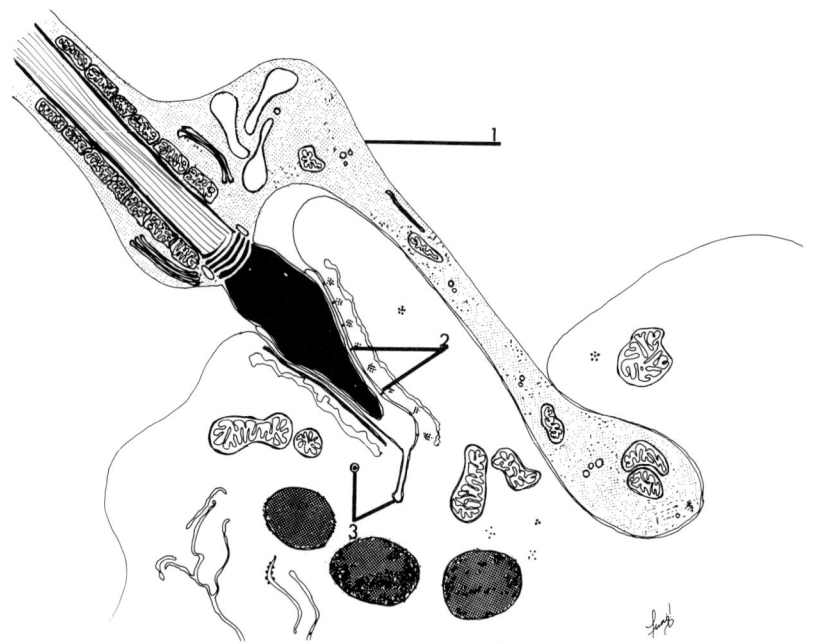

Figure 3.8. This diagram depicts a late elongate spermatid protruding into the lumen of a seminiferous tubule. The spermatid is held in place by three types of anchoring devices: (1) an arm of cytoplasm (stippled) extending from the neck region of the spermatid deep into a recess of the Sertoli cell; (2) septate junctions located between the acrosome region of the spermatid and the surrounding Sertoli cell; and (3) a tubular projection of plasma membrane from the acrosome region of the spermatid, shown in longitudinal and cross section. This idealized diagram is not drawn to scale. The intercellular space at the septate junctions is grossly exaggerated in order to show their structure.

MEMBRANE INTERACTION BETWEEN SERTOLI CELLS AND SPERMATIDS DURING SPERMIOGENESIS

Spermatids are attached to Sertoli cells by intercellular junctions as well as by cytoplasmic interdigitations. Spermatocytes and spermatids are attached by desmosomes and septate junctions to Sertoli cells (Connell, unpublished results; Connell, 1975, 1977; Russell, 1977). These intercellular junctions maintain intercellular adhesion so that an orderly movement of the germ cells from their basal location to their luminal release can occur. In addition to holding the clones of spermatids in the epithelium, however, septate junctions may function to maintain a highly specific cell-to-cell relationship of the haploid cells to specialized regions of Sertoli cell cytoplasm during the process of spermiogenesis. The cytoplasmic interdigitations, the arm or holdfast of cytoplasm retained within a deep recess of Sertoli cell cytoplasm, and the narrow tubular projections of plasma membrane from the acrosome region, help retain the elongate spermatid within the seminiferous tubules during the final stages of maturation (Figure 3.8).

REFERENCES

Bellvé, A. R., J. C. Cavicchia, C. F. Millette, D. A. O'Brien, Y. M. Bhatnagar and M. Dym. "Spermatogenic Cells of Prepubertal Mouse," *J. Cell Biol.* 74:68 (1977).

Black, V. H. "Gonocytes in Fetal Guinea-Pig Testes: Phagocytosis of Degenerating Gonocytes by Sertoli Cells," *Am. J. Anat.* 131:415 (1971).

Carr, J., E. J. Clegg and G. A. Meek. "Sertoli Cells as Phagocytes. An Electron Microscope Study," *J. Anat.* 102:501 (1968).

Chowdhury, A. K. "Thymidine-^{3}H Labeling of Spermatogonia in Rat and Human Seminiferous Tubules Mounted In Toto," *Anat. Record* 169:296 (1971).

Clermont, Y. "Contractile Elements in the Limiting Membrane of the Seminiferous Tubule of the Rat," *Exp. Cell Res.* 15:438 (1958).

Clermont, Y. "The Cycle of the Seminiferous Epithelium in Man," *Am. J. Anat.* 112:35 (1963).

Clermont, Y. "Kinetics of Spermatogenesis in Mammals: Seminiferous Epithelium Cycle and Spermatogonia Renewal," *Physiol. Rev.* 52:198 (1972).

Clermont, Y. and C. Huckins. "Microscopic Anatomy of the Sex Cords and Seminiferous Tubules in Growing and Adult Male Albino Rats," *Am. J. Anat.* 108:79 (1961).

Connell, C. J. "A SEM Study of Spermiogenesis in the Canine Testis," *J. Cell Biol.* 67:78a (1975).

Connell, C. J. "A Freeze-Fracture and Lanthanum Tracer Study of the Complex Junction between Sertoli Cells of the Canine Testis," *J. Cell Biol.* (1978).

Connell, C. J. and G. M. Connell. "The Interstitial Tissue," in *The Testis*, A. D. Johnson and W. R. Gomes, Eds., Vol. IV, Chapter 10 (1977).

Courot, M., M. Hochereau-de-Reviers and R. Ortavant. "Spermatogenesis," in *The Testis*, A. D. Johnson, W. R. Gomes and N. L. Vandemark, Eds., Vol. I (New York: Academic Press, 1970), pp. 273-286.

Dorrington, J. H. and D. T. Armstrong. "Follicle-Stimulating Hormone Stimulates Estradiol-17β Synthesis in Cultured Sertoli Cells," *Proc. Natl. Acad. Sci. USA* 72:2677 (1975).

Dym, M. and D. W. Fawcett. "The Blood-Testis Barrier in the Rat and the Physiological Compartmentation of the Seminiferous Epithelium," *Biol. Reprod.* 3:308 (1970).

Elftman, H. "Sertoli Cells and Testis Structure," *Am. J. Anat.* 113:25 (1963).

Fakunding, J. L., D. J. Tindall, J. R. Dedman, C. R. Mena and A. R. Means. "Biochemical Actions of Follicle-Stimulating Hormone in the Sertoli Cell of the Rat Testis," *Endocrinology* 98:1392 (1976).

Fawcett, D. W. "Ultrastructure and Function of the Sertoli Cell," in *Handbook of Physiology; Endocrinology: Male Reproductive System*, D. W. Hamilton and R. O. Greep, Eds. (Washington: American Physiological Society, 1975), pp. 21-55.

Fawcett, D. W., L. V. Leak and P. M. Heidger. "Electron Microscopic Observations on the Structural Components of the Blood-Testis Barrier," *J. Reprod. Fert. Suppl.* 10:105 (1970).

Franchimont, P., S. Chari and A. Demoulin. "Hypothalamus-Pituitary-Testis Interaction," *J. Reprod. Fert.* 44:335 (1975).

French, F. S. and E. M. Ritzen. "A High Affinity Androgen-Binding Protein (ABP) in Rat Testis: Evidence for Secretion into Efferent Duct Fluid and Absorption by Epididymis," *Endocrinology* 93:88 (1973).

Gondos, B. "Testicular Development," in *The Testis*, A. D. Johnson and W. R. Gomes, Eds., Vol. IV (in press).

Heller, C. G. and Y. Clermont. "Spermatogenesis in Man: An Estimate of its Duration," *Science* 140:184 (1963).

Johnson, F. P. "Dissection of Human Seminiferous Tubules," *Anat. Record* 59:187 (1934).

Josso, N., M. G. Forest and J. Picard. "Müllerian-Inhibiting Activity of Calf Fetal Testis: Relationship to Testosterone and Protein Synthesis," *Biol. Reprod.* 13:163 (1975).

Lacy, D. "Light and Electron Microscopy and its Use in the Study of Factors Influencing Spermatogenesis in the Rat," *J. Royal Microsc. Soc.* 79:209 (1960).

Lacy, D., B. Lofts, B. Kinson, D. Hopkins and H. Dorr. "Sertoli Cells and Steroid Synthesis," *Gen. Comp. Endocrinol.* 5:693A (1967).

Leblond, C. P. and Y. Clermont. "Definition of the Stages of the Cycle of the Seminiferous Epithelium in the Rat," *Ann. N.Y. Acad. Sci.* 55: 548 (1952).

Liang, D. S. "Anatomical Structure of the Testicular Tubules," *Invest. Urol.* 4:285 (1966).

Mancini, R. E., J. L. de la Torré, M. J. Perez del Cerro and O. Vilar. "Histophysiological Aspects of the Sertoli Cell," *Anat. Record* 145:336A (1963).

Mancini, R. E., O. Vilar, B. Alvarez and A. C. Seegreer. "Extravascular and Intratubular Diffusion of Labeled Serum Proteins in the Rat Testis," *J. Histochem. Cytochem.* 13:376 (1965).

Millette, C. F. and A. R. Bellvé. "Temporal Expression of Membrane Antigens during Mouse Spermatogenesis," *J. Cell Biol.* 74:86 (1977).

Moens, P. B. and V. L. W. Go. "Intercellular Bridges and Division Patterns of Rat Spermatogonia," *Z. Zellforsch.* 127:201 (1972).

Nagano, T. "Some Observations on the Fine Structure of the Sertoli Cell in the Human Testis," *Z. Zellforsch.* 73:89 (1966).

Oakberg, E. F. and C. Huckins. "Spermatogonial Stem Cell Renewal in the Mouse as Revealed by ^{3}H-Thymidine Labeling and Irradiation," in *Stem Cells of Renewing Cell Populations*, A. B. Cairnie, P. K. Lala and D. G. Osmond, Eds. (New York: Academic Press, 1976), pp. 287-302.

Perey, B., Y. Clermont and C. P. Leblond. "The Wave of the Seminiferous Epithelium in the Rat," *Am. J. Anat.* 108:47 (1961).

Phillips, D. M. *Spermiogenesis* (New York: Academic Press, 1974).

Roosen-Runge, E. C. "Motions of the Seminiferous Tubules of the Rat and Dog," *Anat. Record* 109:413A (1951).

Roosen-Runge, E. C. "Untersuchungen uber die Regeneration samen bildener Zellen in der normalen Spermatogenese der Ratte," *Z. Zellforsch. Mikroskop. Anat.* 41:221 (1955).

Roosen-Runge, E. C. "Kinetics of Spermatogenesis in Mammals," *Ann. N.Y. Acad. Sci.* 55:574 (1952).

Russell, L. "Desmosome-like Junctions between Sertoli and Germ Cells in the Rat Testis," *Am. J. Anat.* 148:301 (1977).

Russell, L. and Y. Clermont. "Anchoring Device between Sertoli Cells and Late Spermatids in Rat Seminiferous Tubules," *Anat. Record* 185:259 (1976).

Setchell, B. P. and G. M. H. Waites. "The Blood-Testis Barrier," in *Handbook of Physiology; Endocrinology: Male Reproductive System*, D. W. Hamilton and R. O. Greep, Eds. (Washington: American Physiological Society, 1975), pp. 143-172.

Steinberger, A. and E. Steinberger. "Replication Pattern of Sertoli Cells in Maturing Rat Testis *In Vivo* and in Organ Culture," *Biol. Reprod.* 4:84 (1971a).

Steinberger, A. and E. Steinberger. "Hormonal Control of Mammalian Spermatogenesis," *Physiol. Rev.* 51:139 (1971b).

Steinberger, E. "Maturation of Male Germinal Epithelium," in *The Control of the Onset of Puberty*, M. M. Grumbach, G. D. Grave and F. E. Mayer, Eds. (New York: John Wiley & Sons, 1974), pp. 386-402.

Steinberger, E. and A. Steinberger. "Testis: Basic and Clinical Aspects," in *Reproductive Biology*, H. Balin and S. Glasser, Eds. (Amsterdam: Excerpta Medica, 1972), pp. 144-267.

Steinberger, E. and A. Steinberger. "Spermatogenic Function of the Testis," in *Handbook of Physiology; Endocrinology: Male Reproductive System*, D. W. Hamilton and R. O. Greep, Eds. (Washington: American Physiological Society, 1975), pp. 1-19.

Steinberger, E. and D. Y. Tjioe. "A Method for Quantitative Analysis of Human Seminiferous Epithelium," *Fertil. Steril.* 19:960 (1968).

Van der Vusse, G. J., M. L. Kalkman and H. J. Van der Molen. "Endogenous Steroid Production in Cellular and Subcellular Fractions of Rat Testis after Prolonged Treatment with Gonadotropins," *Biochim. Biophys. Acta.* 380:473 (1975).

Witschi, E. "Embryogenesis of the Adrenal and the Reproductive Glands," *Recent Progr. Hormone Res.* 6:1 (1951).

Zaneveld, L. J. D. "The Acrosome of Mammalian Spermatozoa," in *Scanning Electron Microscopic Atlas of Mammalian Reproduction*, E. S. E. Hafez, Ed. (New York: Springer-Verlag, 1975), pp. 58-67.

CHAPTER 4

SPERMATOZOA

E. S. E. Hafez

The human testis is extremely sensitive to changes in its environment. This sensitivity can be evaluated with remarkable precision in the cellular components of the ejaculate. Human semen is unique when compared with other mammalian semen. For example, there is a high percentage of morphologically abnormal spermatozoa in specimens from donors of high fertility. Also the standard, oval-shaped sperm head is not uniform in size or shape. Because of this lack of uniformity in human spermatozoa, several rating schemes for morphologic characteristics have been described.

The length of normal spermatozoa is variable ranging from 2.5-10 μm, with an average of 5 μm (Figure 4.1). Normal human spermatozoa are characterized by flattened and ovoid heads covered anteriorly by a rough, rigid surface and posteriorly by a smooth surface with few granules. The surface of the posterior portion of the head is smoother than that of the anterior portion. The heads of human spermatozoa appear pointed at the posterior end and became somewhat round and flat anteriorly (Hafez *et al.*, 1975). Elevations appeared in the region between the anterior and posterior portions, and a shallow furrow was observed. There are remarkable individual variations in the shape of this furrow. The size, shape and internal structure of the acrosome are subject to modification as is the organization of the mitochondria of the middle piece. A transverse furrow is observed on the head.

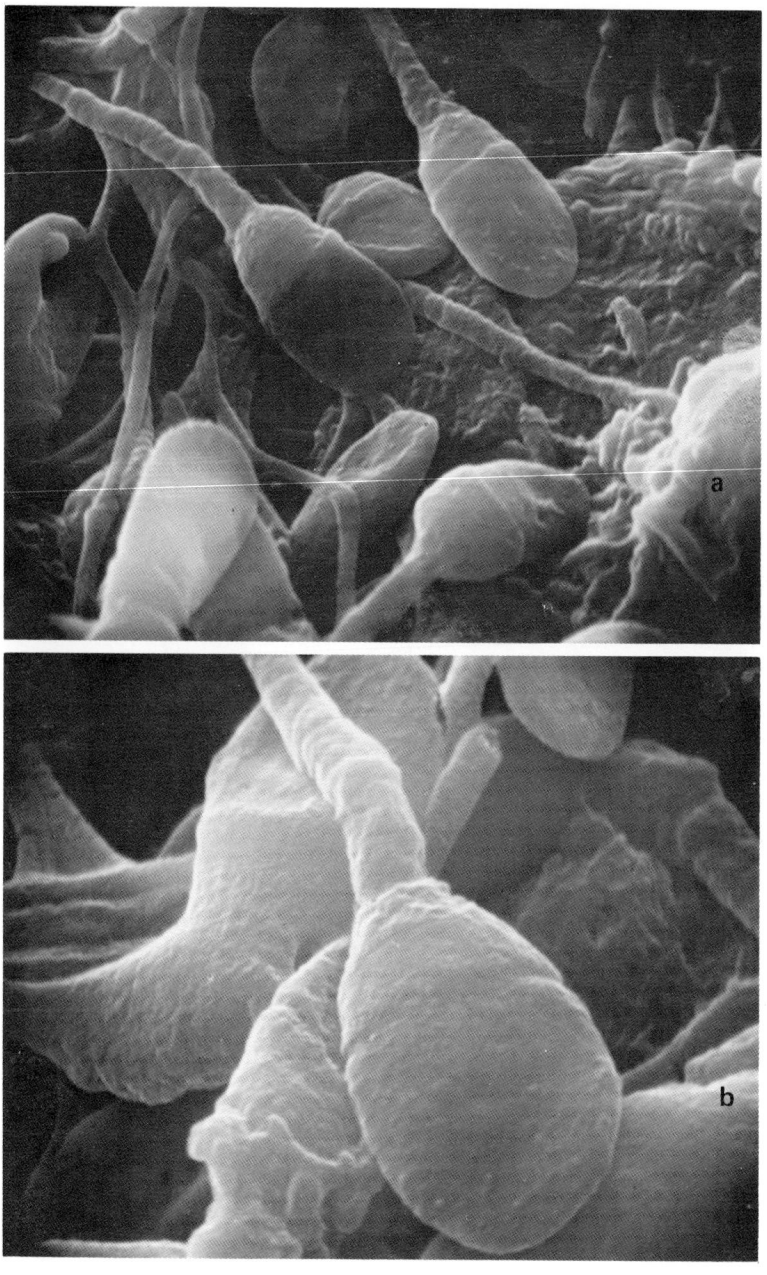

Figure 4.1. Human spermatozoa after coagulation and liquefaction. Note heterogenity of size and shape of head and acrosome; (a) X1000; (b) X1400 (photos by J. E. Flechon and E. S. E. Hafez).

There are considerable variations in the morphology and thickness of the neck of spermatozoa. In some spermatozoa, the neck appears as a thickening or as a slight constriction connecting the midpiece and tail, which apparently is due to the gap between mitochondrial and fibrous sheaths (Hafez et al., 1975). The tail is surrounded by a piece of fibrous tail sheath. The endpiece is characterized by an abrupt narrowing.

ABNORMAL SPERMATOZOA

Abnormal spermatozoa are observed in all human ejaculates. The percentage of abnormal spermatozoa in average ejaculates is observed to be 20-30% higher under the scanning electron microscope than with light microscopy. Abnormalities of the head include large, deformed, spheroid formation or release of the cytoplasm near the head, forming a cyst-like structure. Amorphous spermatozoa have structural defects in the shape or size of the head. Oval, large, small or tapering, or bicephalic forms are also common.

Several types of gross abnormality of the sperm head have been observed with the light microscope, and examples of some of these have been observed with scanning electron microscopy (Figures 4.2, 4.3 and 4.4). With the exception of a few sperm which displayed ruptured membranes at the apex of the acrosome, most of the more extreme variations in head shape are due to a malformation of the nucleus rather than the acrosome cap. The large lumpy projection on the head may be caused by vacuolation in the cytoplasm or by evagination of the acrosome. Bicephalia of spermatozoa is common in the presence of varicocele. Duplicated rudimentary heads, unusually small, and deformed heads with multiple tails, are also observed. The midpiece in duplicated spermatozoa is thick and shows a longitudinal furrow suggesting fusion of a pair of midpieces. The spermatids which appear in the ejaculate are usually mononuclear but often are found in clusters sharing common cytoplasm. In men of known fertility and otherwise normal seminal quality, the dominant spermatozoa is the "oval" form.

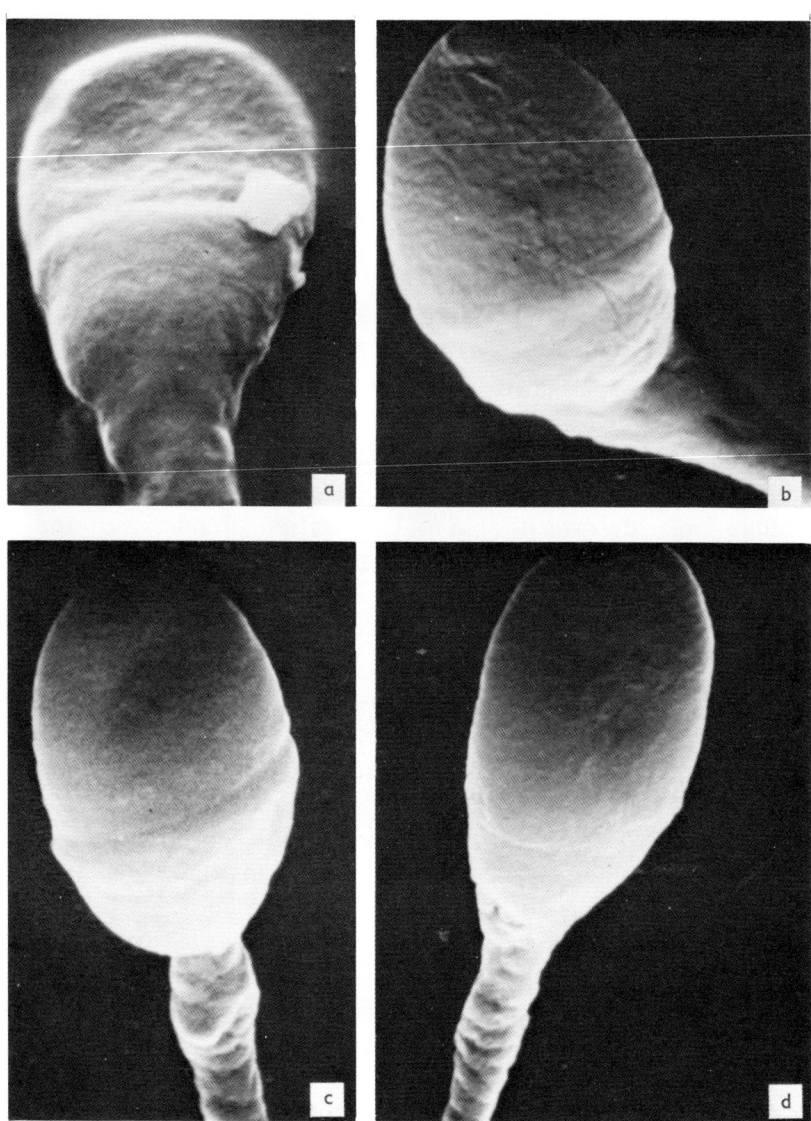

Figure 4.2. Spermatozoa from fertile men. Note differences in morphology of sperm head: (a) X13,000; (b) X13,000; (c) X13,000; (d) X12,000 (Hafez *et al.*, 1975; photos by J. Flechon).

Andrology 61

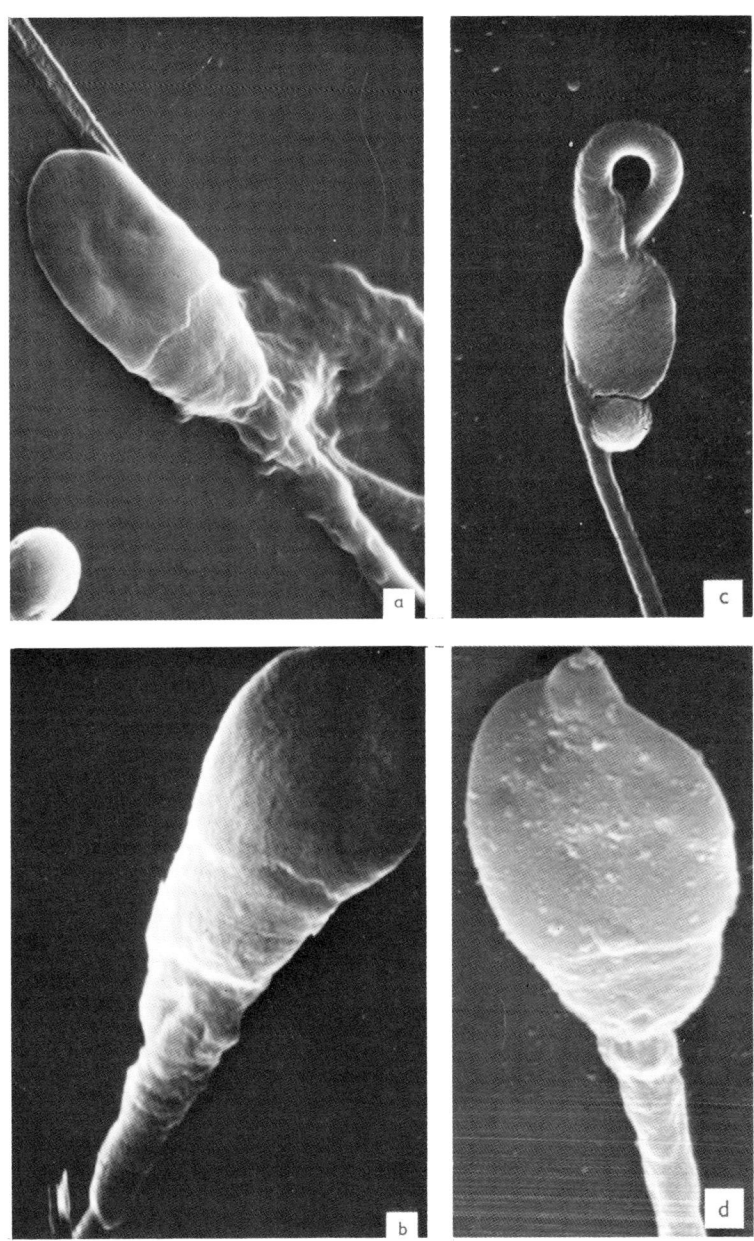

Figure 4.3. Abnormal human spermatozoa: (a) X10,800; (b) X11,700; (c) X7800; (d) X10,800 (photos by E. S. E. Hafez and J. E. Flechon).

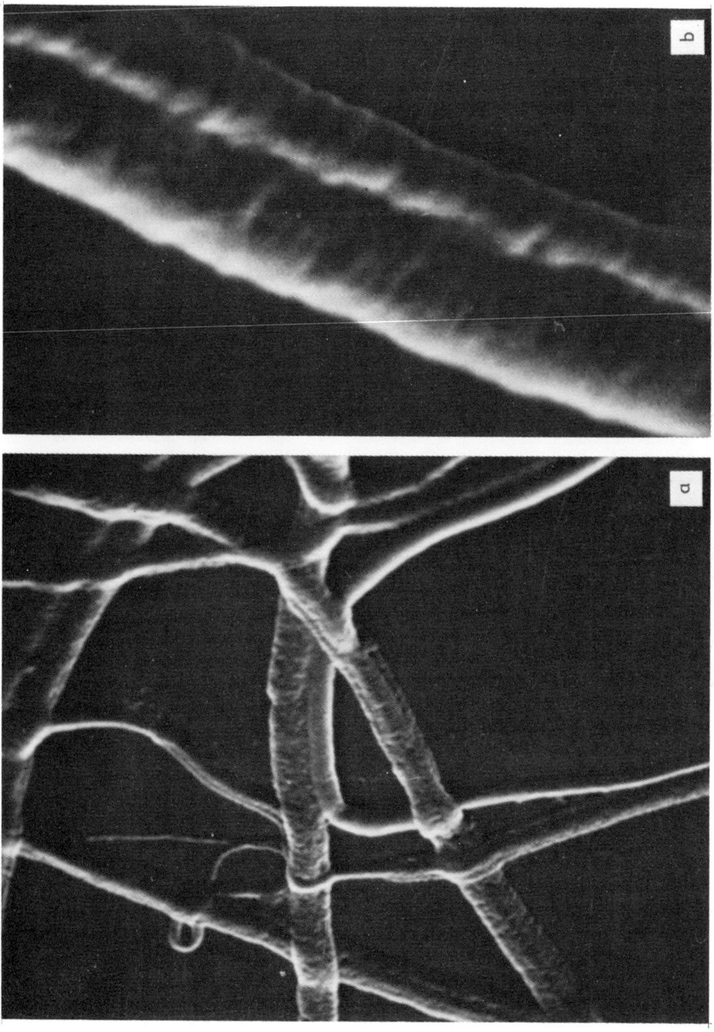

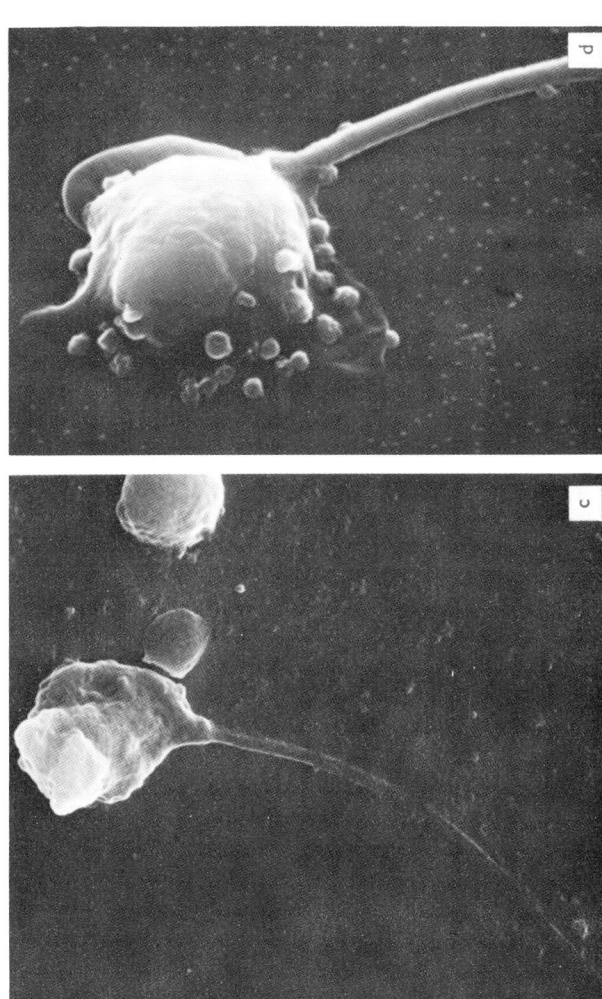

Figure 4.4. (a) Tails of human spermatozoa (X12,000). (b) Duplicated human sperm tail (X60,000). (c) Phagocytosis of spermatozoa by leukocytes (photos by E. S. E. Hafez and J. E. Flechon).

Abnormalities of the midpiece included parts of thin and constricted midpieces, enlargement, breakage and duplication. The morphological anomalies may be a result of trauma, illness, or the use of antispermatogenic compounds (Hafez et al., 1975).

PHAGOCYTOSIS OF SPERMATOZOA

Phagocytosis of spermatozoa by leukocytes is not uncommon; phagocytosis is one of the most striking properties of the plasma membrane. The nucleus and cytoplasm of the leukocyte covers most parts of the sperm head. The cytoplasm of the leukocyte engulfs the sperm head. This seems to represent phagocytosis of sperm by leukocytes in the female reproductive tract. The phagocytic sequence begins when the spermatozoan becomes attached to the outside surface of a leukocyte, although the mode of spermatozoan attachment is variable. One such mode involves contact between the anterior region of the spermatozoan head and the leukocyte (Hafez et al., 1975). Thereafter, the leukocyte surface membrane creeps up around the edge of the spermatozoan head, forming a gullet-like tube of membrane which ultimately engulfs the head of the spermatozoan. During ingestion, the leukocyte plasmalemma remains closely apposed to the plasma membrane of the spermatozoan head but does not exert undue compression, since the head membrane remains intact.

FUNCTIONAL SIGNIFICANCE

The transverse furrow observed on the head of human spermatozoa under SEM seems to correspond to the light zone observed under light microscopy and to the infolding of plasma membrane in the gap between the acrosome and postnuclear cap (Fujita et al., 1970; Pedersen, 1969). The apical segment of the acrosome is relatively small, simple in form, and extends but little beyond the top of the nucleus (Fawcett and Phillips, 1969). The plasma membrane of the acrosome is irregularly undulated and partly lifted. Swelling and vesiculation of the acrosome as described by Pedersen (1969) may be due to physiologic changes or to artifacts (Hadek, 1969). The human spermatozoan, unlike that of many other mammals, has no subacrosomal space or "perfortiosum" (Bedford, 1967). There is no convincing evidence for the existence of an acrosomal cap or galea capitis in human spermatozoa. In other mammals, spermatozoa have well-defined acrosomes or head caps. It is possible that in human spermatozoa an acrosome has been present, but that it is lost in a very early stage before or during ejaculation, when the spermatozoa are admixed with the seminal plasma, which in man is rather alkaline and shows high proteolytic activity (Van Duijn, 1967).

The acrosome contains proteolytic enzymes (Stanbaugh and Buckley, 1970; Yanagimachi and Noda, 1972) which probably facilitate sperm penetration in cervical mucus and in luminal fluids in the uterus and the zona pellucida. Little is known about the possibility of the mammalian subacrosomal substance playing the same important role in fertilization as does the corresponding cytoplasm in some invertebrate types (Nicander and Bane, 1967).

Head-to-head agglutination is observed in certain patients. Little is known about physiological and pathological factors which cause agglutination of spermatozoa. Sperm agglutination results from hemagglutinating myxoviruses (Peleg, 1966), Sendai virus (Buthala et al., 1971) and Mycoplasma species (Taylor-Robinson and Manohee, 1967).

The detached apical cytoplasmic substances are observed on the cell surfaces in the oviduct and uterus. The oviductal fluid plays a major role in the transport and maturation of the gametes (Fredricsson and Bjorkman, 1962). There is a close similarity between the oviductal fluid and serum contents (immunoglobulins, electrolytes, glucose and enzymes), indicating that serum transudation is an important contributory source of oviductal secretions (Lippes et al., 1972). At about the time of expected ovulation, the lymphatic vessels, especially those in the lamina propria of the free fimbrial "villi," appeared greatly distended resembling a labyrinthine-like channel system (Ferenczy and Richart, 1974). In estrogen-deprived postmenopausal patients, the oviductal epithelium is characterized by deciliation and loss of secretory activity. These changes occur in all segments of the oviduct and only rarely are there clusters of cilia, which appear short, scant and often degenerating (Ferenczy and Richart, 1974).

REFERENCES

Bedford, J. M. "Observations on the Fine Structure of Spermatozoa of the Bush Baby (*Galagos senegalensis*), the African Green Monkey (*Cercopithecus aethiops*) and Man," *Am. J. Anat.* 121:443 (1967).

Buthala, D. A., R. J. Ericsson and G. T. Chubb. "Interaction of Sendai Virus and Rabbit Sperm: Transmission and Scanning Electron Microscopy," *Biol. Reprod.* 5:325 (1971).

Fawcett, D. W. and D. M. Phillips. "Recent Observation of the Ultrastructure and Development of the Mammalian Spermatozoan," in *Comparative Spermatology* (Rome: Academia Nationale Dei Lincei, 1969).

Ferenczy, A. and R. M. Richart. "Female Reproductive System," in *Dynamics of Scanning and Transmission Electron Microscopy* (New York: John Wiley & Sons, 1974), p. 213.

Fredricsson, B. and N. Bjorkman. "Studies on the Ultrastructure of the Human Oviduct Epithelium in Different Functional States," *Z. Zellforsch. Mikrosk. Anat.* 58:387 (1962).

Fujita, T., M. Miyoshi and J. Tokunaga. "Scanning and Transmission Electron Microscopy of Human Ejaculate Spermatozoa with Special Reference to their Abnormal Forms," *Z. Zellforsch. Mikrosk. Anat.* 105:483 (1970).

Hadek, R. "Mammalian Fertilization," in *An Atlas of Ultrastructure* (New York: Academic Press, 1969).

Hafez, E. S. E., M. I. Barnhart, H. Ludwig, J. Lusher, I. Joelsson, J. L. Daniel, A. I. Sherman, J. A. Jordan, H. Wolf, W. C. Stewart and F. C. Chretien. "Scanning Electron Microscopy of Human Reproductive Physiology," *Acta Obstet. Gynecol. Scand.*, Suppl. 40 (1975).

Lippes, J., R. G. Enders, D. A. Pragay and W. R. Bartholomew. "The Collection and Analysis of Human Fallopian Tubal Fluid," *Contraception* 5:85 (1972).

Nicander, L. and A. Bane. "An Electron Microscopical Study of the Acrosomes of some Mammalian Spermatozoa," *Fertil. Steril.* (New York: Excerpta Medica, 1967).

Pedersen, H. "Ultrastructure of the Ejaculated Human Sperm," in *Comparative Spermatology* (Rome: Academia Nationale dei Lincei, 1969).

Peleg, B. A. and P. Ianconescu. "Sperm Agglutination and Serum Absorption due to Myxoviruses," *Nature (London)* 211 (1966).

Stambaugh, R. and J. Buckley. "Comparative Studies of the Acrosomal Enzymes of Rabbit, Rhesus Monkey and Human Spermatozoa," *Biol. Reprod.* 2:275 (1970).

Taylor-Robinson, D. and R. J. Manohee. "Sperm Adsorption and Sperm Agglutination by Mycoplasms," *Nature (London)* 215:484 (1967).

Van Duijn, C., Jr. "Cytological Structure of Human Spermatozoa, Revealed by Optical Microscopy," *Fertil. Steril.* (1967).

Yanagimachi, R. and Y. D. Noda. "Scanning Electron Microscopy of Golden Hamster Spermatozoa before and during Fertilization," *Experientia* 28:68 (1972).

CHAPTER 5

THE SEMINAL COAGULUM

P. F. Tauber, D. Propping and L. J. D. Zaneveld

THE PHENOMENON OF SEMEN COAGULATION AND LIQUEFACTION

Human semen is ejaculated as a viscous, yellowish-gray, opaque fluid that coagulates during or immediately after ejaculation. The coagulum appears as an amorphous, elastic, sticky clot and has gel-like properties. Shortly thereafter, usually within 5-20 minutes, the solid parts of the gel undergo spontaneous liquefaction and the ejaculate becomes liquid again. Both coagulation and liquefaction are presumed to be enzymatic in nature and occur *in vitro* as well as in the vagina (Huggins and Neal, 1942; Oettle, 1954; Sobrero and MacLeod, 1962). The rate and extent of liquefaction seem to be directly related to the structural aspects of the seminal coagulum (Oettle, 1954; Zaneveld *et al.*, 1974a). Impairment or complete lack of semen coagulation and liquefaction due to seminal vesicle or prostate pathology are at times a cause of male subfertility (Amelar, 1962; Amelar and Hotchkiss, 1963; Bunge, 1970; Bunge and Sherman, 1954; Huggins, 1945; Moon and Bunge, 1968).

SPECIES DIFFERENCES

The property of human semen to coagulate and liquefy within a certain time after emission can be considered unique. Semen from nonhuman primates also coagulates and liquefies partially, but the majority remains as coagulum both *in vitro* and in the vagina (Hoskins and Patterson, 1967). Semen from some small rodents, such as the rat and guinea pig, coagulates instantaneously after ejaculation and forms a firm clot (vaginal plug) that fails to liquefy even after many days. By contrast, the ejaculate of the dog and the bull does not coagulate at all (Tauber and Zaneveld, 1976).

PHYSIOLOGICAL IMPORTANCE

The importance of coagulation in human fertility remains to be established. By contrast, the function of the seminal plug in rodents can be explained from a mechanical standpoint. In these animals, spermatozoa are ejaculated ahead of the seminal plug. The plug forces the spermatozoa in close contact with the cervix and also prevents the backflow of semen from the vagina.

The entire human ejaculate is coagulated and thus consists of a mixture of spermatozoa and accessory sex gland secretions. Most spermatozoa are completely entrapped in the protein meshwork of the coagulum and cannot escape unless liquefaction occurs. The spermatozoa are therefore effectively immobilized during the time between ejaculation and liquefaction. From a teleological standpoint, one can speculate that this *motility gap* between ejaculation and liquefaction is a resting phase for the spermatozoa that protects their energy supply which is later needed for the migration to the ovum (MacLeod, 1952). Additionally, if the coagulum is located near the cervical os, gradual liquefaction allows spermatozoa to enter the cervix immediately without having to contact vaginal fluid. The acidic pH of human vaginal fluid (approximately 4.0) may otherwise be detrimental to sperm motility.

COAGULUM FORMATION
AND LIQUEFACTION (Table 5.1)

In some animals, such as the rat, the guinea pig and the rhesus monkey, the coagulum is formed by the secretions from the cranial or anterior lobe of the prostate (Tauber and Zaneveld, 1976; Van Wagenen, 1936). This lobe is also called the "coagulating gland" and, though embryologically of the same origin, differs somewhat in morphology from the rest of the prostate. Coagulum formation in rodents requires *vesiculase*, an enzyme that derives almost solely from the coagulating gland (Williams-Ashman et al., 1972). The action of vesiculase is not species-specific since mixing the prostatic secretion of one species (rat) with seminal vesicle secretion of a different species (monkey) or vice versa results in coagulation (Van Wagenen, 1936). Biochemically, vesiculase acts as a transamidase that causes protein cross-linking in the coagulable substrate, which in turn leads to the formation of an extremely firm clot (Williams-Ashman et al., 1972). This action of vesiculase is therefore similar to that of the activated factor XIII in mammalian blood coagulation (Lorand, 1965).

Neither a "coagulating gland" nor an enzyme similar to vesiculase have been found in man. Using the split ejaculation technique where the semen is collected in consecutive portions, it can be shown that coagulation usually does not occur in the first portion of such a split ejaculate. If it occurs, liquefaction takes place almost immediately (Tauber et al., 1973). By contrast, the last portion forms a firm coagulum that liquefies rather slowly, usually over a period of 20-30 minutes or longer. Addition of the first portion to the last portion enhances the speed of liquefaction of the last portion (Zaneveld et al., 1974b). Based on the concentration of various biochemical components in such split ejaculates (Tauber et al., 1975; Tauber et al., 1976), it is known that the first portion mainly consists of prostatic secretion whereas the last portion chiefly represents secretions from the seminal vesicles. Thus, the coagulable substrate originates from the seminal vesicles, and the liquefying agent derives from the prostate gland. Neither the absolute number nor the integrity of the spermatozoa present in the ejaculate influences

Table 5.1
Morphology of Mammalian Semen Coagula

Species	Formation of Semen Coagulum	Occurrence of Liquefaction	Location of Spermatozoa	Appearance of Fibers	Thickness of Fibers (μm)	Ultrastructural Appearance of Coagulum	Possible Physiological Effect/Function
Rodent, Guinea pig	Yes	No	Outside Coagulum	Short, flat	0.6	Sponge-like, rigid, dense	Vaginal plug, prevention of backflow of spermatozoa
Subhuman Primate, Rhesus monkey	Yes	Partially (approx. 25%)	Mostly outside coagulum but also within	Variable, thick	0.2-0.7	Solid, disorganized array of fibers	Same as guinea pig
Human, Normal, before liquefaction	Yes	Yes, complete	Trapped inside the coagulum	Long, narrow	0.15	Dense, organized network of fibers	Not known
Human, Normal, during liquefaction	—	—	—	Fibers turn into spherical globules	0.15-3.0 (globules)	Globules	Required for release of spermatozoa
Human, Slowly liquefying sample before liquefaction	Yes	Yes, but much slower than normal (1-2 hr)	Trapped inside the coagulum	Large, with network of small fibers	0.8-1.8	Thick and solid	Clinically associated with male subfertility

the liquefying activity (Von Kaulla and Shettles, 1953; Tauber *et al.*, 1973; Zaneveld *et al.*, 1974b). Coagulation and liquefaction disorders of human semen may therefore occur regardless of whether the donor is normospermic, oligospermic or azoospermic.

The seminal coagulum of man resembles morphologically a fibrin-blood clot. This observation and the fact that the human seminal coagulum has the property to liquefy, have led to the suggestion that a "sperm fibrin" is present in semen that is in turn dissolved by a fibrinolytic enzyme (Lundquist, 1952). Although human semen possesses plasminogen activator activity, neither fibrin nor fibrinogen are present (Tauber *et al.*, 1975). Other components involved in blood clot formation and lysis are also mostly or entirely absent from human semen such as prothrombin, factors XII and XIII and plasminogen (Huggins and Neal, 1942; Tauber and Zaneveld, 1976). Therefore, the coagulation and liquefaction of human semen is completely dissimilar to that of blood. Some indications are present that the liquefaction of the human seminal coagulum is due to a prostatic proteinase, called seminin (Syner *et al.*, 1975; Tauber and Zaneveld, 1976).

MORPHOLOGY OF THE HUMAN COAGULUM

Using the light microscope, the solid portions of the seminal coagulum show a variety of interlacing bundles of clearly defined, parallel "refractile" fibers. During liquefaction, this fibrous texture loses continuity and contact. The fibers become irregular, break up and finally disappear totally. By then, all fibrous material has turned into spherical material and the remaining fluid appears floccular. Finally, only liquid seminal plasma is found.

More detailed information regarding the structural aspects of the seminal coagulum before and during liquefaction can be obtained with the SEM. Immediately after ejaculation, the human coagulum represents a large amorphous mass that consists of long fibrous strands (Figure 5.1). A closer look reveals that the surface of the clot exhibits a tightly organized network of short and narrow fibers, approximately 0.15 μ in diameter that are layered in bundles and cross at regular intervals. Only very small

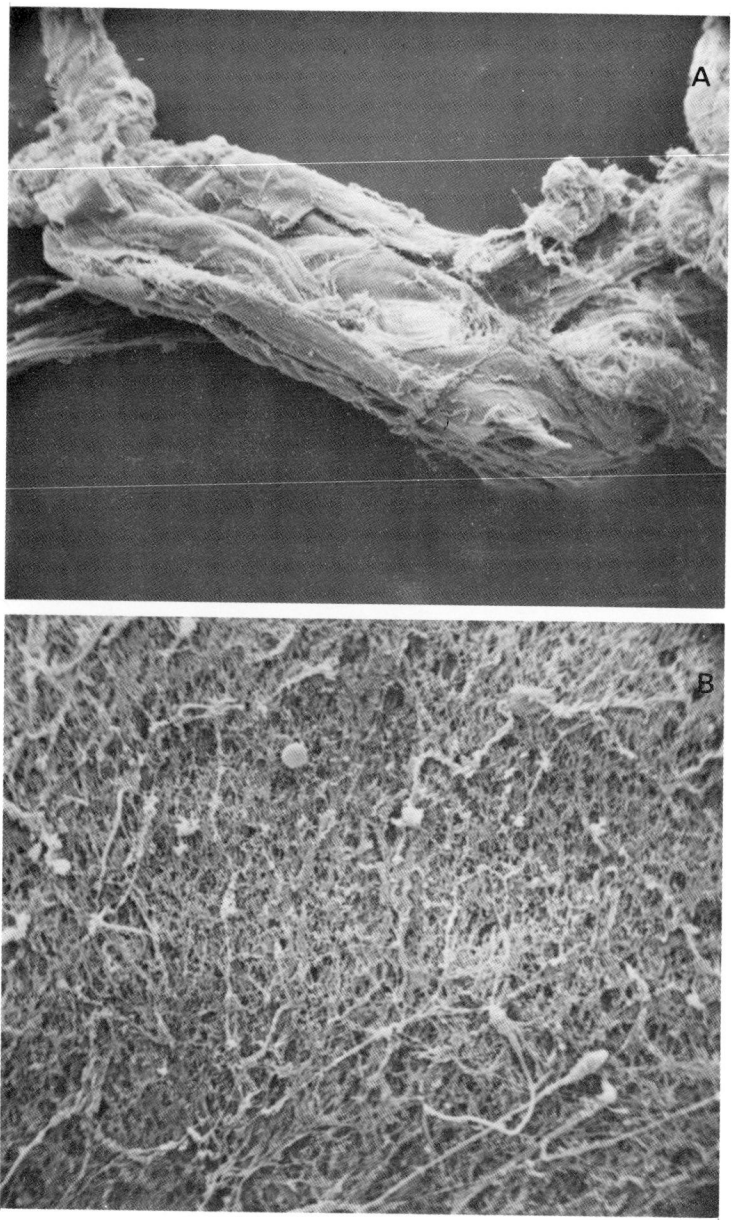

Figure 5.1. Human seminal coagulum 3 minutes after ejaculation. A-C: normal coagulum. D: slow lysing coagulum. (A) X30; (B) X600;

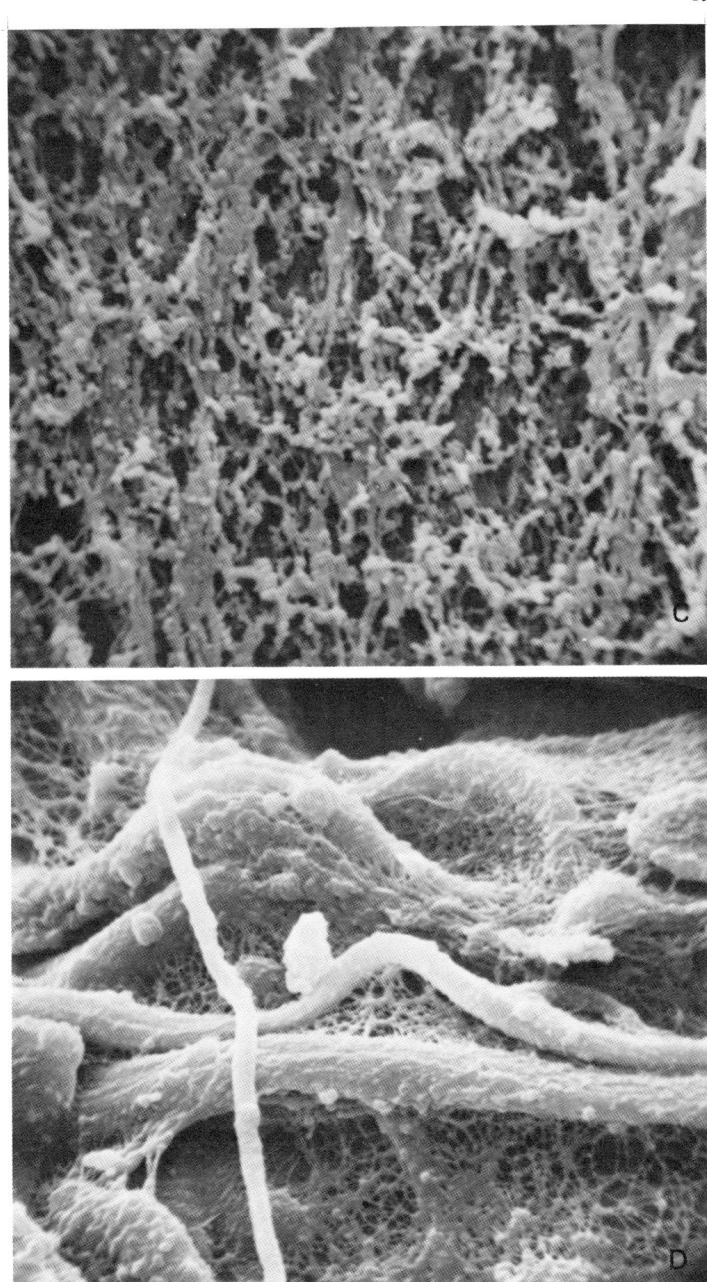

(C) X3000; (D) X2875 (Zaneveld *et al.*, 1974a).

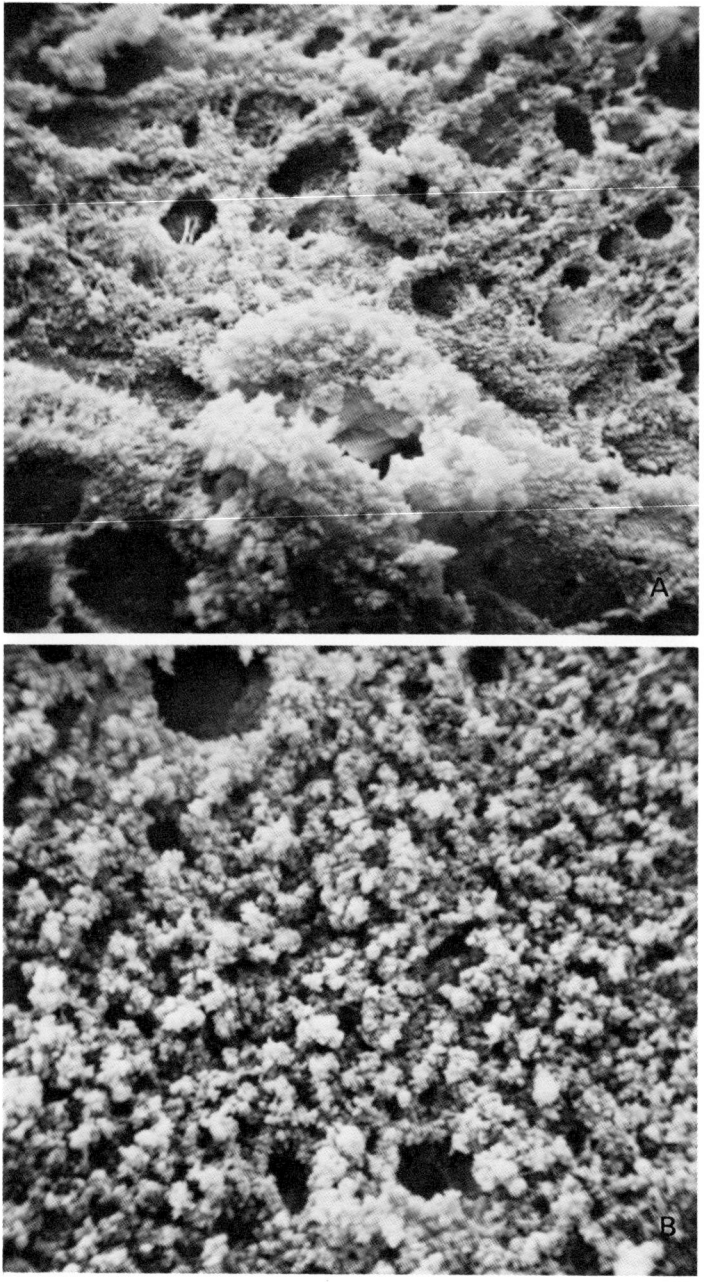

Figure 5.2. Liquefying human seminal coagulum. A and B: partially liquefied, 6 minutes after ejaculation. C and D: completely liquefied,

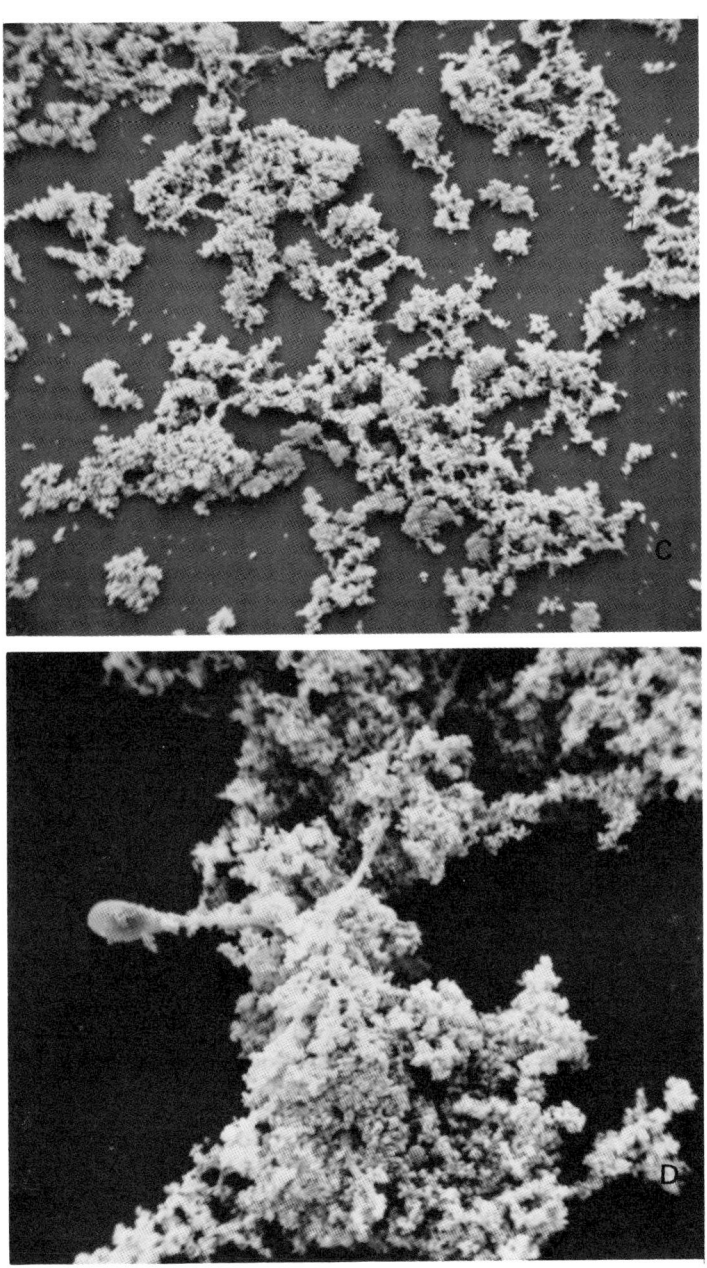

15 minutes after ejaculation. (A) X1200; (B) X3100; (C) X1250; (D) X1200 (Zaneveld et al., 1974a).

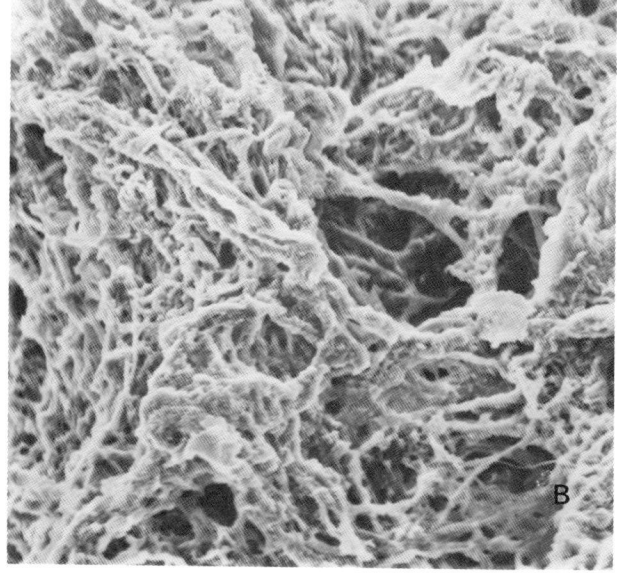

Figure 5.3. Human seminal coagulum collected from the human vagina.

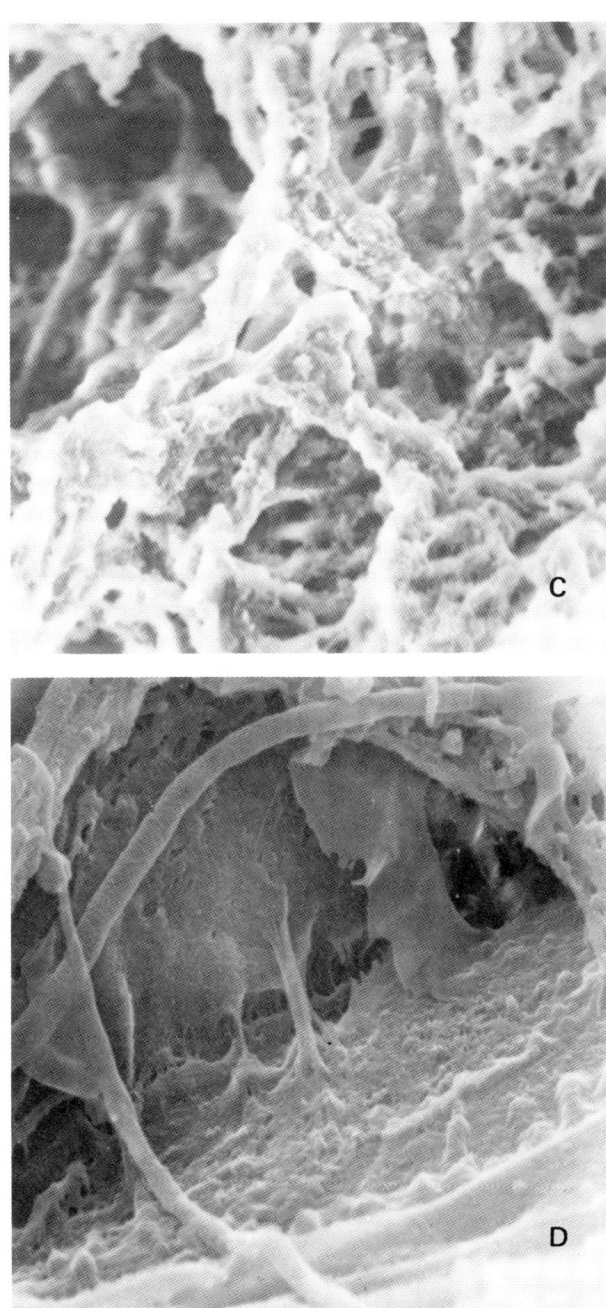

(A) X520; (B) X500; (C) X1250; (D) X6000.

spaces are left between these fibers—too small to allow movement of entrapped spermatozoa that are dispersed throughout the fibrous arrangement.

At the start of liquefaction, amorphous material becomes visible (Figure 5.2). This material turns into small spherical globules, and the spaces between the fibers become larger. The coagulum retains its orginal shape for a time, but as more of the fibers disappear and the globules take over, the coagulum loses its structure. Usually after about 5-20 minutes, depending on the speed of liquefaction, only globules can be found. These globules either occur separately or are grouped in clumps. Occasionally, spermatozoa are partially trapped in such a clump.

Distinct structural differences are present between a normal lysing coagulum and a semen sample with slow liquefaction (Figure 5.1D). The slowly lysing ejaculate possesses large fibers that measure approximately 1.0-2.0 μ in diameter (Zaneveld et al., 1974a). However, the general arrangement of the fibrous texture does not differ from a normal ejaculate. The large fibers are connected to each other and to a dense meshwork of fine, small fibers, similar in size to those of a normal lysing coagulum. More time is probably required to liquefy these large fibers, thus explaining the retardation of liquefaction.

The speed of lysis of a slowly liquefying coagulum can be enhanced and even brought to normal when the semen is deposited in the vagina. The morphological appearance of the *in vivo* process is similar to that *in vitro* (Figure 5.3). The fibrous texture begins to lyse as early as 5 minutes after deposition of semen in the vagina. The spaces between the fibers increase and spherical, globule-like material occurs at the surface. The large fibers become thinner and finally disrupt. It is as yet not known how the vaginal environment increases the rate of liquefaction.

After being released from the coagulum, the spermatozoon can be studied with the SEM (Figure 5.4). It is approximately 50 μ long, with a head that is 5 μ long, 3 μ wide and 2 μ thick, and a midpiece that is about 5 μ long. The anterior half of the human sperm head is flat whereas it is rounded at its posterior end. Various grooves can be found on the posterior portion of the sperm head. One of these is particularly prominent in some spermatozoa and may be the posterior boundary of the acrosome.

Andrology 81

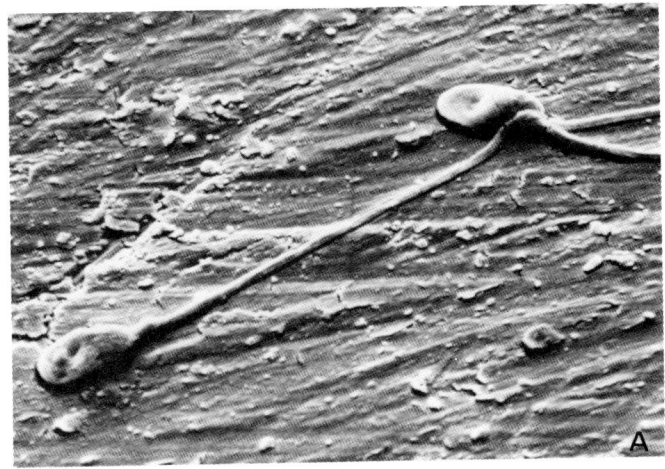

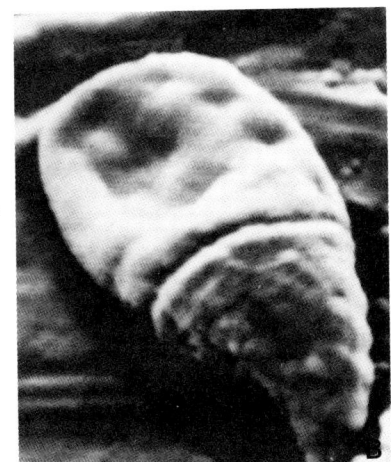

Figure 5.4. Human spermatozoa. (A) X2000; (B) X11,000 (Zaneveld *et al.*, 1971).

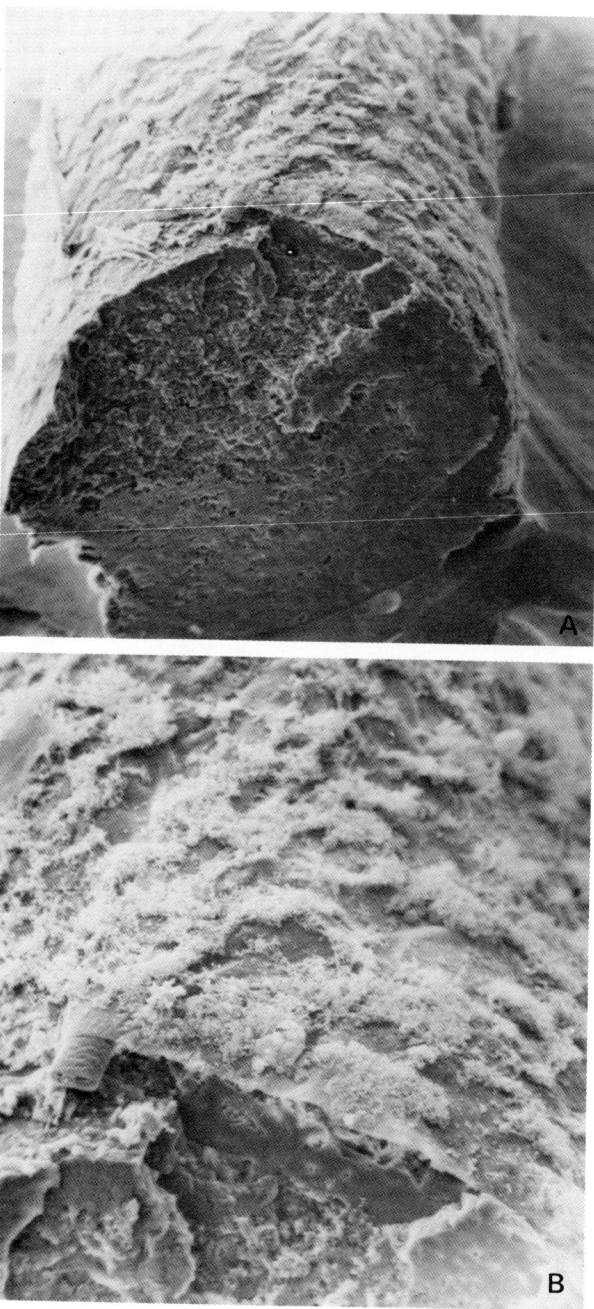

Figure 5.5. Guinea pig coagulum. (A) X16; (B) X40; (C) X1600;

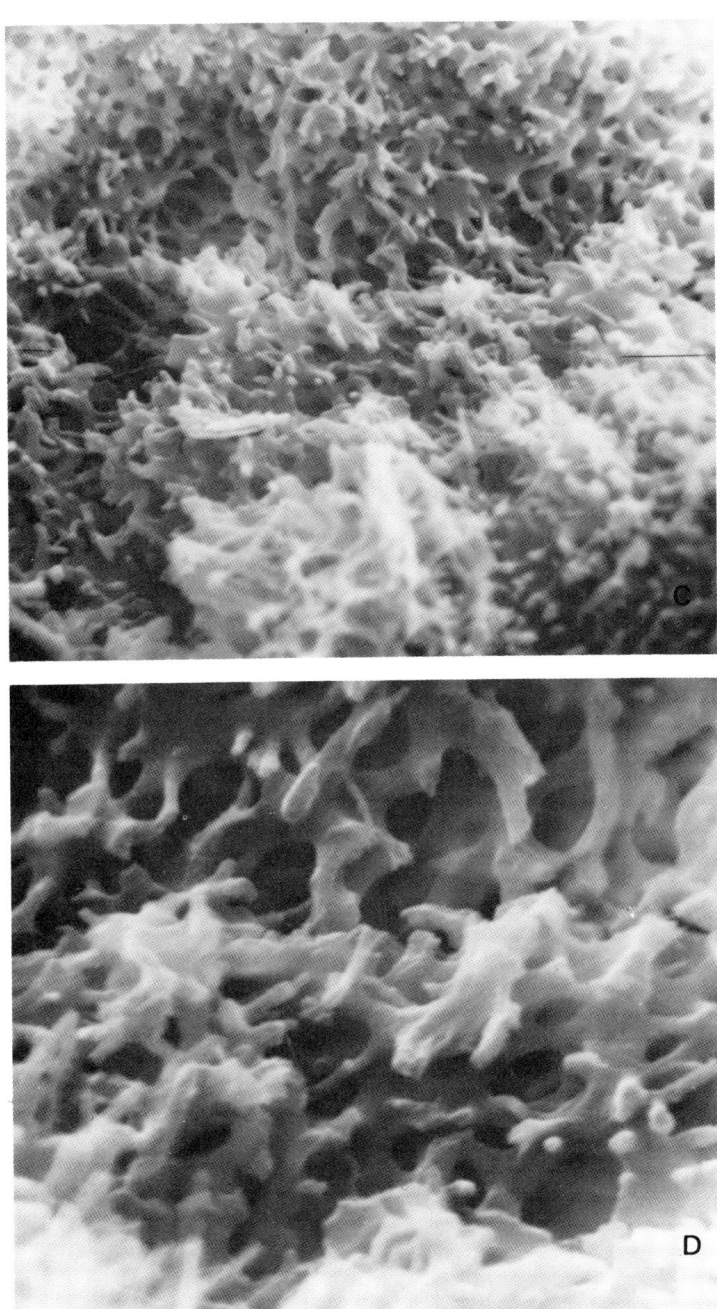

(D) X4000 (Zaneveld *et al.*, 1974a).

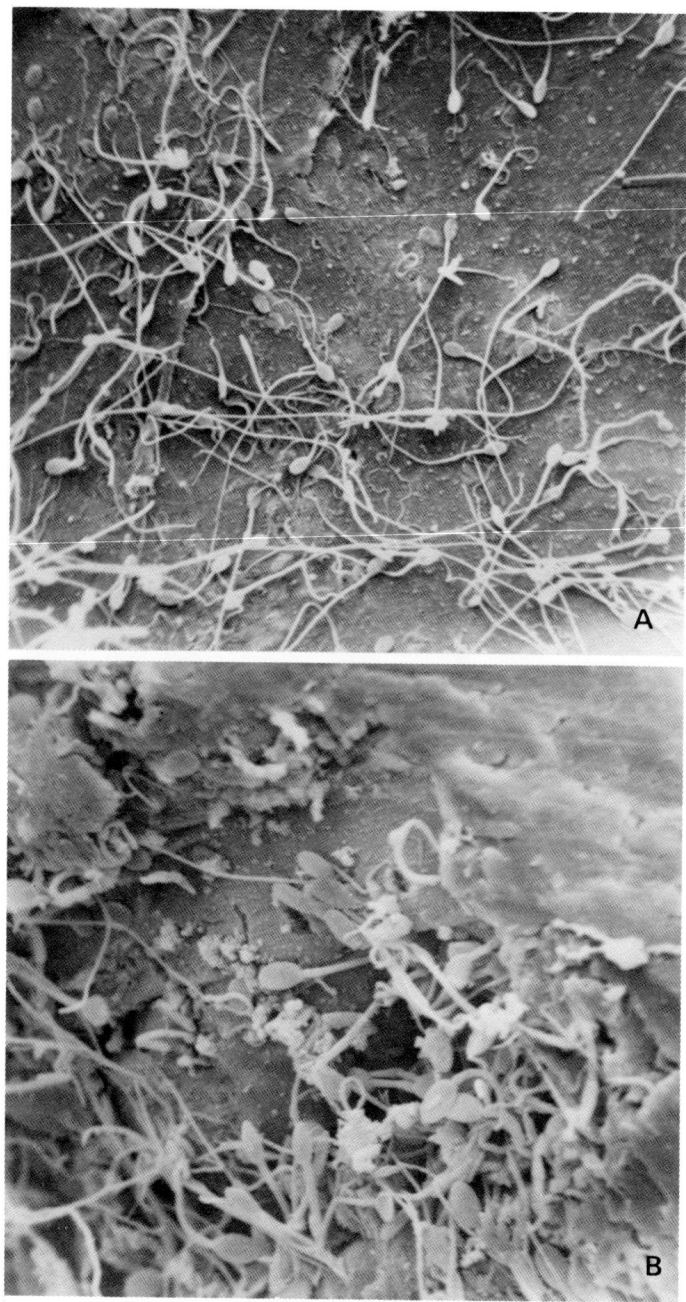

Figure 5.6. Rhesus monkey coagulum. (A) X310; (B) X660; (C) X130;

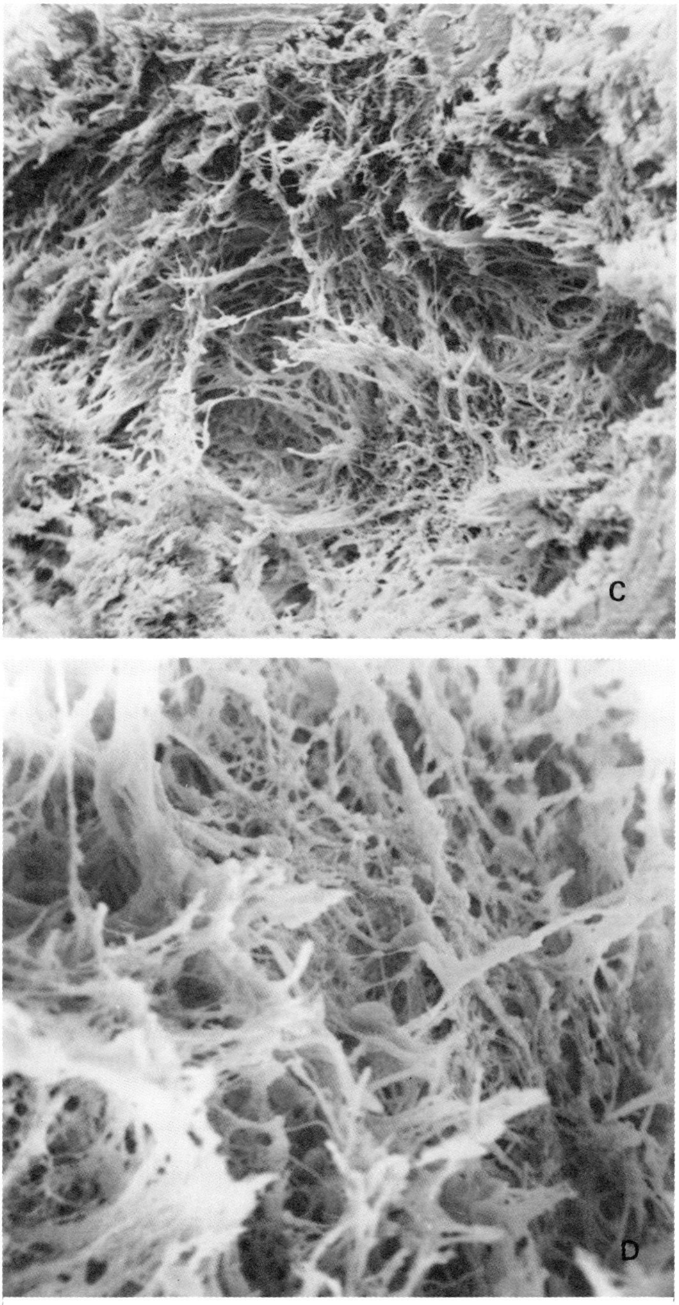

(D) X650 (Zaneveld *et al.*, 1974a).

With the exception of abnormally shaped spermatozoa, most spermatozoa appear to be the same, and there is no evidence that X-spermatozoa differ in morphology from Y-spermatozoa.

MORPHOLOGY OF ANIMAL COAGULA

Significant differences can be detected when the human coagulum is compared with the seminal clot of the guinea pig and the rhesus monkey using the SEM (Zaneveld et al., 1974a). The guinea pig plug represents a rigid, tubular structure that on close observation shows a texture similar to a "sponge" (Figure 5.5), consisting of short and flat fibers. These fibers are approximately 0.6 μ thick, about the same size as those of the monkey coagulum. This surface of the plug has a smooth appearance and is partially covered by flaky material. No spermatozoa can be found inside or on the surface of the plug.

The rhesus monkey coagulum is much more similar to that of the human than to that of the guinea pig (Figure 5.6). The surface of a fractured specimen exhibits the presence of long fibers of variable thickness (0.2-0.7 μ); about 2-5 times larger than the fibers in a human coagulum. The fibers are also more irregularly arranged than those of the human coagulum and the whole clot appears more solid. Some spermatozoa can be found within the fibrous texture but most of them are present on the surface. Many spermatozoa remain on the surface even after extensive washing. Some adhesive properties may therefore exist between the monkey coagulum and the spermatozoa.

ACKNOWLEDGMENTS

The authors wish to thank Dr. G. F. B. Schumacher, in whose laboratory many of the previously published SEM studies were performed, and Ms. L. B. Roberts for the typing of this manuscript. The unpublished portions of this research were supported by NIH grant HD 09868.

REFERENCES

Amelar, R. D. "Coagulation, Liquefaction and Viscosity of Human Semen," *J. Urol.* 87:187 (1962).

Amelar, R. D. and R. S. Hotchkiss. "Congenital Aplasia of the Epididymides and Vasa Deferentia: Effects on Semen," *Fertil. Steril.* 14:44 (1963).

Bunge, R. G. "Some Observations on the Male Ejaculate," *Fertil. Steril.* 21:639 (1970).

Bunge, R. G. and J. K. Sherman. "Liquefaction of Human Semen by Alpha-amylase," *Fertil. Steril.* 5:353 (1954).

Hoskins, D. D. and D. L. Patterson. "Prevention of Coagulum Formation with Recovery of Motile Spermatozoa from Rhesus Monkey Semen," *J. Reprod. Fertil.* 13:337 (1967).

Huggins, C. "The Physiology of the Prostate Gland," *Physiol. Rev.* 25:281 (1945).

Huggins, C. and W. Neal. "Coagulation and Liquefaction of Semen. Proteolytic Enzymes and Citrate in Prostatic Fluid," *J. Exp. Med.* 76:527 (1942).

Lorand, L. "Physiological Roles of Fibrinogen and Fibrin," *Fed. Proc.* 24:784 (1965).

Lundquest, F. "Studies on the Biochemistry of Human Semen. IV. Amino Acids and Proteolytic Enzymes," *Acta Physiol. Scand.* 25:178 (1952).

MacLeod, J. "Biochemistry of the Male Genital Tract," *Ann. N.Y. Acad. Sci.* 54:796 (1952).

Mann, T. *Biochemistry of Semen and of the Male Reproductive Tract* (New York: John Wiley & Sons, 1965).

Moon, K. H. and R. G. Bunge. "Observations on the Biochemistry of Human Semen. III. Amylase," *Fertil. Steril.* 19:977 (1968).

Oettle, A. G. "Morphologic Changes in Normal Human Semen After Ejaculation," *Fertil. Steril.* 5:227 (1954).

Sobrero, A. J. and J. MacLeod. "The Immediate Post Coital Test," *Fertil. Steril.* 13:184 (1962).

Syner, F. N., K. S. Moghissi and J. Yanez. "Isolation of a Factor from Normal Human Semen that Accelerates Dissolution of Abnormally Liquefying Semen," *Fertil. Steril.* 26:1064 (1975).

Tauber, P. F., D. Propping, L. J. D. Zaneveld and G. F. B. Schumacher. "Biochemical Studies on the Lysis of Human Split Ejaculates," *Biol. Reprod.* 9:62 (1973).

Tauber, P. F., L. J. D. Zaneveld, D. Propping and G. F. B. Schumacher. "Components of Human Split Ejaculates. I. Spermatozoa, Fructose, Immunoglobulins, Albumin, Transferrin, Lactoferrin and Other Plasma Proteins," *J. Reprod. Fertil.* 43:249 (1975).

Tauber, P. F. and L. J. D. Zaneveld. "Coagulation and Liquefaction of Semen," in *Human Semen and Fertility Regulation in Men*, E. S. E. Hafez, Ed. (St. Louis: Mosby, 1976), p. 153.

Tauber, P. F., L. J. D. Zaneveld, D. Propping and G. F. B. Schumacher. "Components of Human Split Ejaculates. II. Enzymes and Proteinase Inhibitors," *J. Reprod. Fertil.* 46:165 (1976).

Van Wagenen, G. "The Coagulating Function of the Cranial Lobe of the Prostate Gland in the Monkey," *Anat. Rec.* 66:411 (1936).

Von Kaulla, K. N. and L. B. Shettles. "Relationship Between Human Seminal Fluid and the Fibrinolytic System," *Proc. Soc. Exp. Biol. Med.* 83:692 (1953).

Williams-Ashman, H. G., A. G. Notides, S. S. Pabalan and L. Lorand. "Transamidase Reactions Involved in the Enzymatic Coagulation of Semen: Isolation of Gamma-Glutamyl-E-Lysine Dipeptide from Clotted Secretion Protein of Guinea Pig Seminal Vesicle," *Proc. Natl. Acad. Sci. U.S.A.* 69:2322 (1972).

Zaneveld, L. J. D., K. G. Gould, W. J. Humphreys and W. L. Williams. "Scanning Electron Microscopy of Mammalian Spermatozoa," *J. Reprod. Med.* 6:152 (1971).

Zaneveld, L. J. D., P. F. Tauber, C. Port, D. Propping and G. F. B. Schumacher. "Scanning Electron Microscopy of the Human, Guinea Pig and Rhesus Monkey Seminal Coagulum," *J. Reprod. Fertil.* 40:223 (1974a).

Zaneveld, L. J. D., G. F. B. Schumacher, P. F. Tauber and D. Propping. "Proteinase Inhibitors and Proteinases of Human Semen," in *Proteinase Inhibitors*, H. Fritz, H. Tschesche, L. J. Greene and E. Truscheit, Eds. (New York: Springer-Verlag, 1974b), p. 136.

CHAPTER 6

THE PROSTATE GLAND

Elinor Spring-Mills and Albert L. Jones

GENERAL STRUCTURE AND FUNCTION

The adult prostate is the largest accessory gland within the male reproductive tract. It is a firm, encapsulated body, about 3.8 cm in diameter, which weighs approximately 20 g. The prostate surrounds the urethra below the internal urethral orifice at the neck of the bladder. During embryonic life, five groups of tubular evaginations from the walls of the primitive posterior (prostatic) urethra merge at their boundaries and give rise to the immature gland. As a result, it has been the custom to subdivide the male prostate into five lobes (Figure 6.1), even though the lobes within the adult gland are continuous and cannot be unequivocally differentiated either macroscopically or microscopically (Huggins, 1945; Lowsley, 1962; McNeal, 1972).

The prostate contains 30-50 compound tubuloalveolar glands which empty into the prostatic urethra via 20-30 excretory ducts. Its moderately dense fibroelastic stroma contains abundant smooth muscle fibers, blood vessels, lymphatics and nerves.

The alkaline secretion of the gland forms the bulk of the seminal fluid. In addition to reducing the acidity of the urethra, it decreases the viscosity of semen and augments sperm motility.

The prostate is neither essential to life nor is its presence an absolute requirement for fertility. It is, however, of major medical importance since benign or malignant tumors of the prostate affect almost every male in the United States by the age of 50.

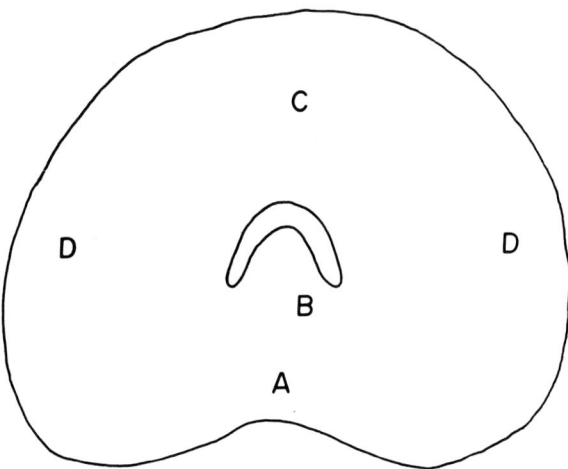

Figure 6.1. Diagram shows a cross section of the human prostate at the level of the colliculus seminalis. The approximate position of the posterior (A), median (B), anterior (C) and two lateral (D) lobes is shown. The lateral and median lobes are the most frequent sites for benign nodular hyperplasia, whereas the posterior lobe is the most common site of prostatic carcinoma.

CAPSULE AND STROMA

The fibroelastic stroma (Figure 6.2), which supports and ensheaths the parenchyma, accounts for approximately 25% of the total volume of the gland. Smooth muscle cells and fibroblasts are the most common cell types. These cells, together with the abundant collagen and elastic fibers, presumably impart the characteristic elastic consistency to the gland, which is used for distinguishing normal from abnormal glands during physical examination. In certain diseases of the prostate, the stroma increases substantially, either alone or in conjunction with the epithelium (Mostofi and Price, 1973; Franks, 1974; Tannebaum, 1975; Webber, 1975).

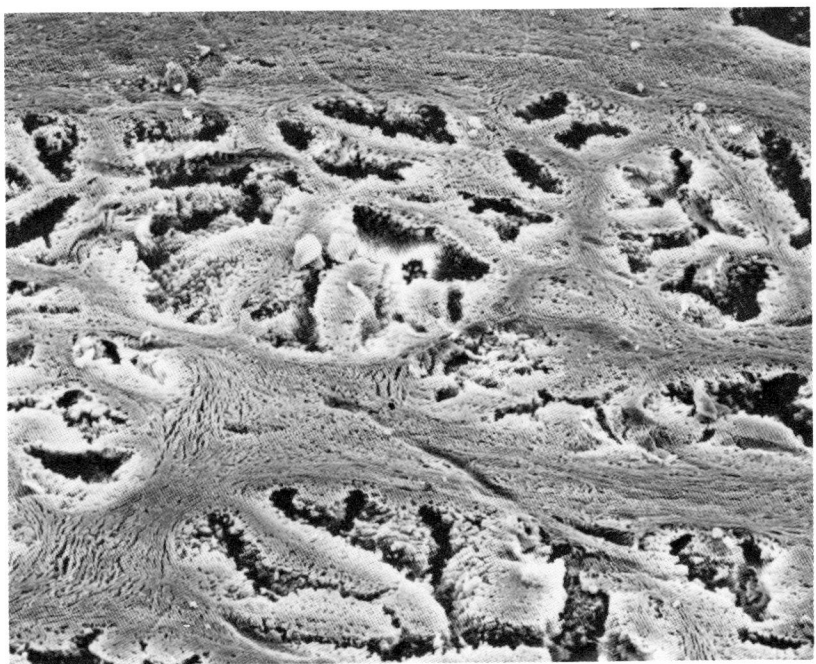

Figure 6.2. The stroma of the prostate contains dense fibroelastic connective tissue and abundant smooth muscle cells. The broad stromal trabeculae, which extend from the capsule into the interior of the gland where they surround tubules and ducts, can be seen in this scanning electron micrograph (X120).

PARENCHYMA

The elongated tubules of the gland normally vary in size and shape (Figures 6.2, 6.3 and 6.4). They are highly irregular, tortuous and branching. Saccular recesses are usually present, and cystic dilations of the alveoli and ducts are common (Figures 6.4 and 6.5).

Within the alveoli of humans and rats, two types of epithelial cells have been described (Brandes, 1966; Fisher and Jeffrey, 1965). The predominant cell is a columnar secretory or glandular epithelial cell containing a moderate number of mitochondria, apical Golgi apparatus and densely packed secretory vacuoles.

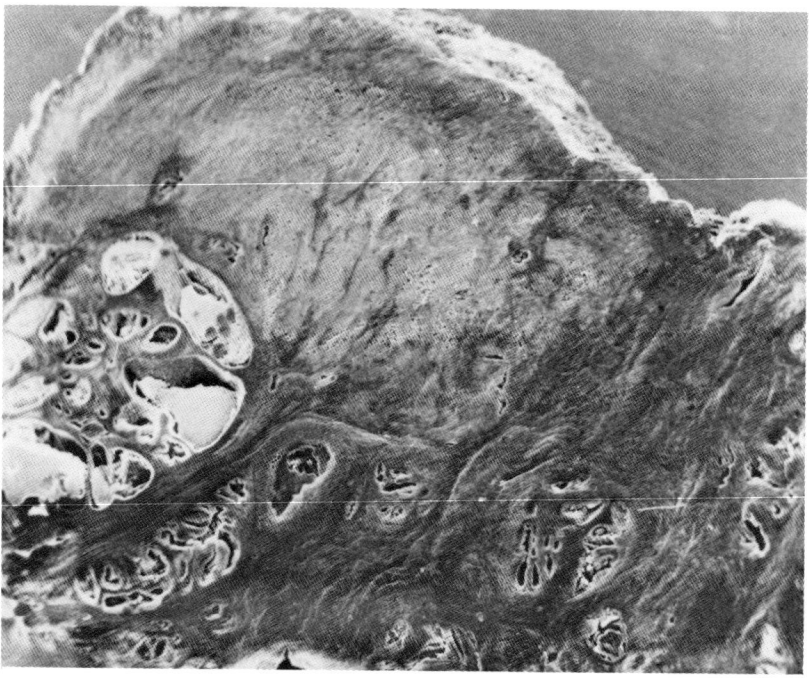

Figure 6.3. Tubules, ducts, stroma and cystically dilated alveoli are present in this specimen from a patient with prostatic carcinoma (X23).

The apical surfaces of these cells bear a moderate coat of short microvilli (Figure 6.6). Basal cells are polygonal in shape. They lack secretory vesicles and contain few mitochondria. Their endoplasmic reticulum, Golgi apparatus and ribosomes also are poorly developed.

SECRETION

The immature prostate is very small, nonpalpable and does not secrete. Maturation of the gland and the maintenance of normal structure and function after puberty depend on the hormonal secretions of the testis. Secretion is continuous in the adult and is especially active during coitus. *Corpora amylacea*, which presumably are condensations of the secretion, increase in

Andrology 93

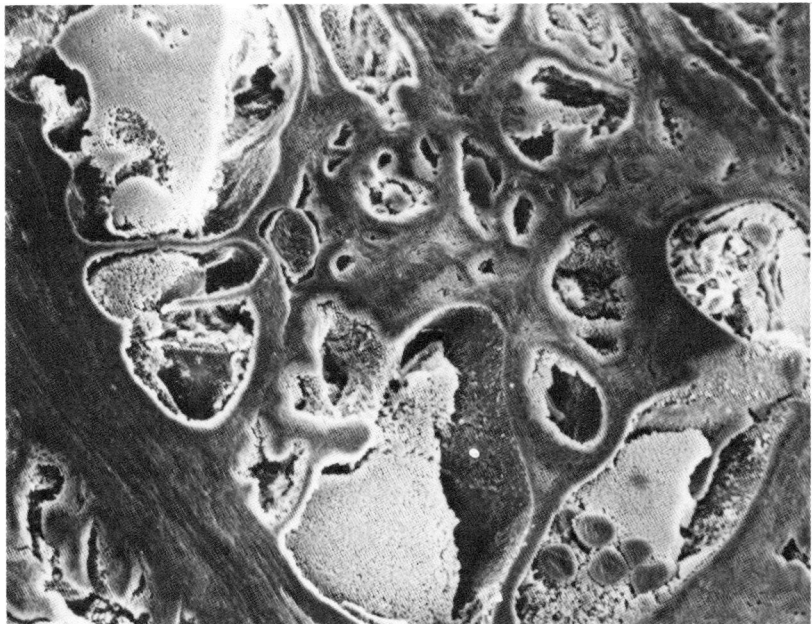

Figure 6.4. The region in Figure 6.3 with cystically dilated alveoli is shown here at higher magnification (X69).

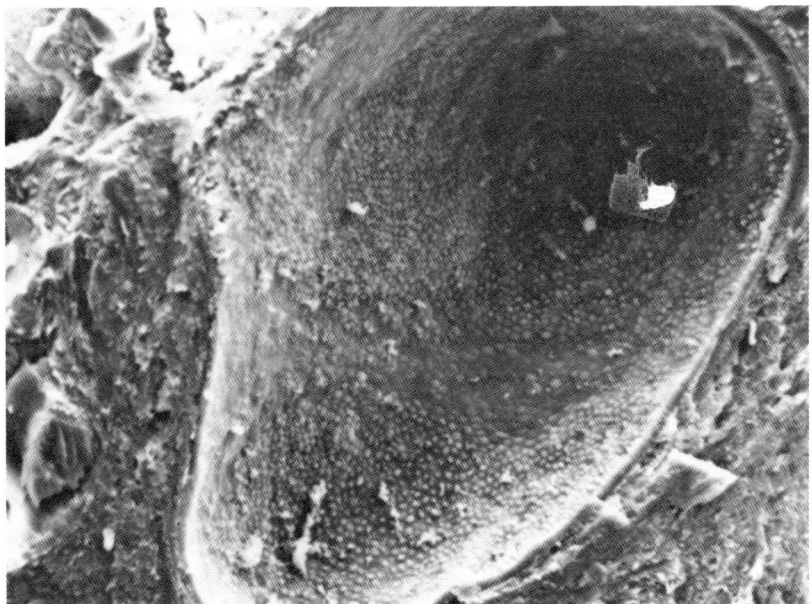

Figure 6.5. A saccular recess of a prostatic tubule is shown. The lumen is lined by columnar epithelial cells. Strands of coagulated secretory material adhere to the cells in several areas (X150).

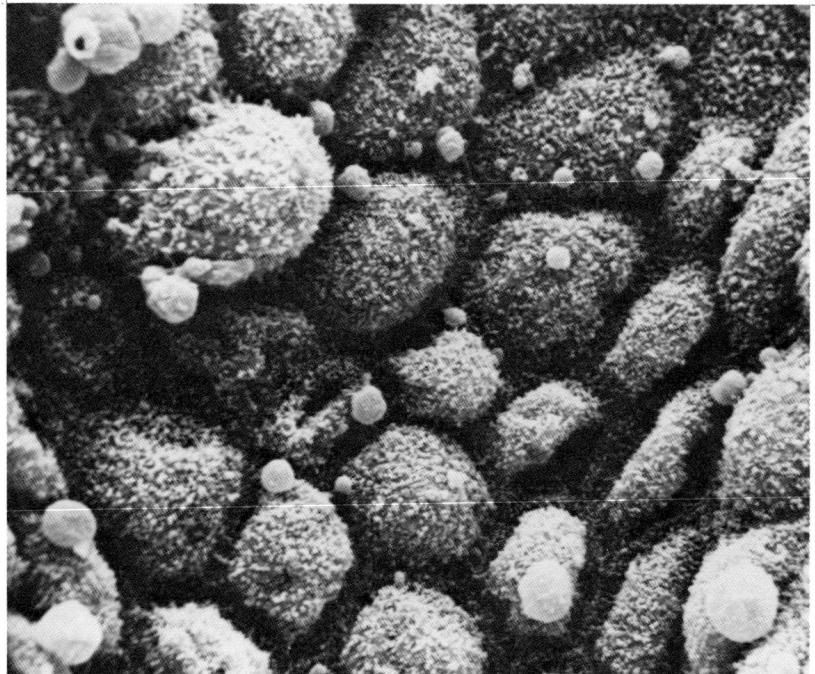

Figure 6.6. Short, stubby microvilli and cytoplasmic blebs, which may be secreted into the lumen, are present on the apical surfaces of the columnar epithelial cells, which form the bulk of the gland's parenchyma (X3625).

number with age and are extremely variable in size. The largest of these concretions, which cannot pass out of the alveoli, are retained within the gland and often calcify (Figure 6.7).

ACKNOWLEDGMENTS

The authors wish to thank Drs. N. Scott McNutt and C. Glassman for providing the prostate specimens; The ETEK Corporation for the use of their SEM Autoscan; Ms. Maria Maglio for specimen preparation; and Mrs. Naomi Ross for excellent secretarial assistance.

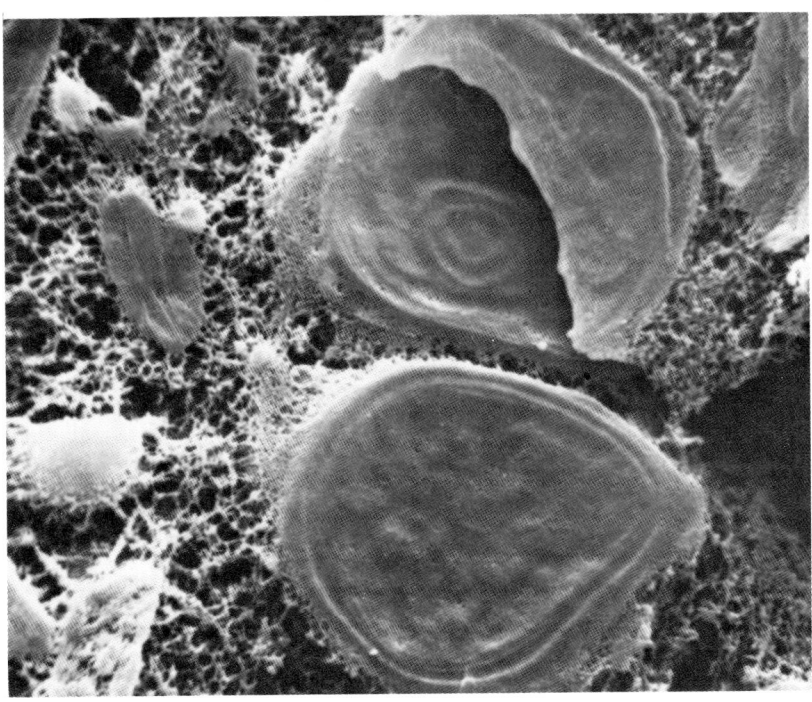

Figure 6.7. Two ovoid, lamellated prostatic concretions are shown in this scanning electron micrograph. Fixation has coagulated the remainder of the secretion into a reticulated mass (X608).

REFERENCES

Brandes, D. "The Fine Structure and Histochemistry of Prostatic Glands in Relation to Sex Hormones," *Int. Rev. Cytol.* 20:207 (1966).

Fisher, E. R. and W. Jeffrey. "Ultrastructure of Human Normal and Neoplastic Prostate: with Comments Relative to Prostatic Effects of Hormonal Stimulation in the Rabbit," *Am. J. Clin. Pathol.* 44:119 (1965).

Franks, L. M. "Biology of the Prostate and its Tumors," in *The Treatment of Prostatic Hypertrophy and Neoplasia*, J. E. Castro, Ed. (Baltimore: University Park Press, 1974), pp. 1-26.

Huggins, C. "The Physiology of the Prostate Gland," *Physiol. Res.* 25:281 (1945).

Lowsley, O. S. "Development of the Human Prostate with Reference to the Development of Other Structures at the Neck of the Bladder," *Am. J. Anat.* 13:299 (1962).

McNeal, J. E. "The Prostate and Prostatic Urethra: A Morphologic Synthesis," *J. Urol.* 107:1008 (1972).

Mostofi, F. K. and E. B. Price, Jr. *Atlas of Tumor Pathology: Tumors of the Male Genital System* (Washington, DC: Armed Forces Institute of Pathology, 1973).

Tannebaum, M. "Diagnostic Criteria for Histopathologic Evaluation of Prostatic Tissue Sections," *Urology* 5:407 (1975).

Webber, M. M. "Ultrastructural Changes in Human Prostatic Epithelium Grown *In Vitro*," *J. Ultrastruc. Res.* 50:89 (1975).

SECTION III

GYNECOLOGY

CHAPTER 7

THE OVARY

S. Makabe and E. S. E. Hafez

The cuboidal germinal epithelial cells, polygonal in shape, are covered with microvilli (Figure 7.1). The cortex of the human ovary is covered with fibroblastic-like cells with numerous microvilli. Stereocilia may be observed among the microvilli. The tunica albuginea is made up of fine and coarse networks of longitudinal fibers. The theca interna cells have a flat superficial pattern and appear like a string of beads (Ludwig and Metzger, 1976).

PREOVULATORY FOLLICLE

Preovulatory follicles are blister-like structures which protrude from the ovarian surface. Follicular cells around the base of preovulatory follicles are made of polyhedral or cuboidal cells with several microvilli (Figure 7.1). Follicular cells located in front of the antrum are flattened due to stretching and possess elongated ameboid evaginations and/or cytoplasmic protrusions with very few microvilli.

In the preovulatory follicle, there is progressive disruption of the follicle wall, superficial epithelium, and the connective tissue of the ovarian cortex (Nilsson and Munski, 1973). In some apical regions of the follicle, cells of the superficial epithelium are flattened, lose their microvilli, and slough off (Motta and Van Blerkom, 1975).

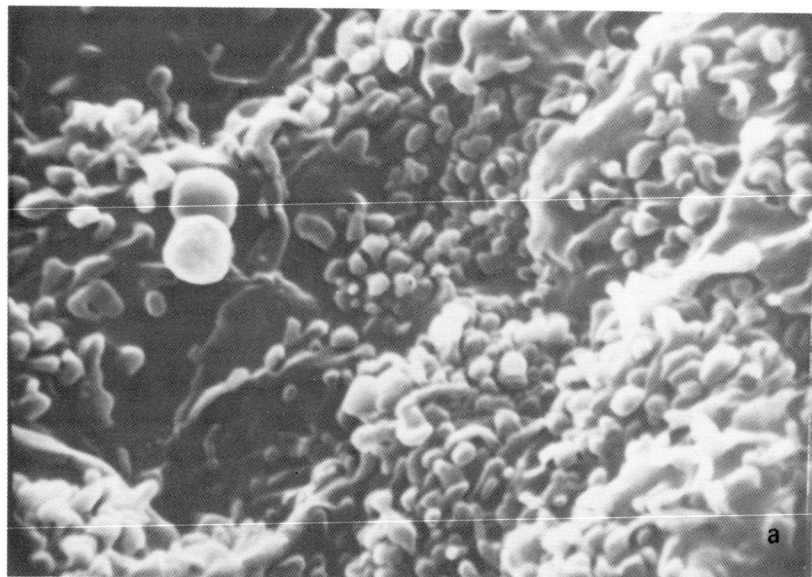

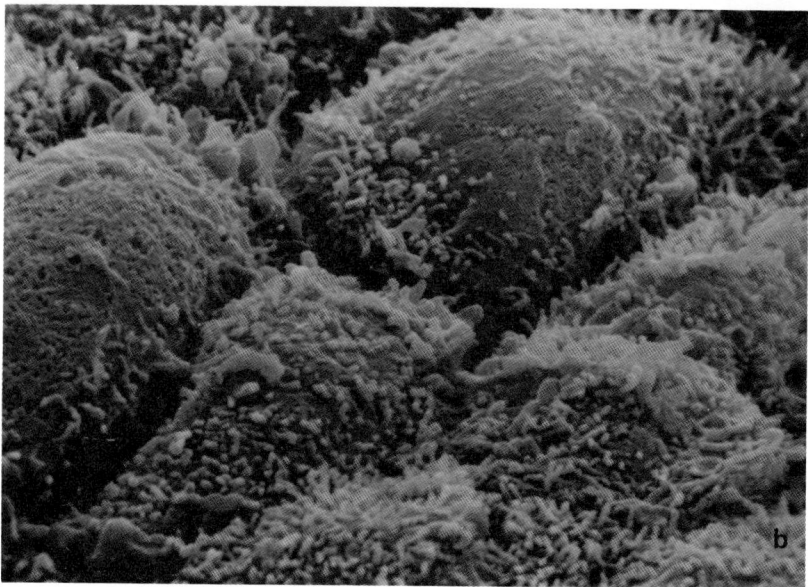

Figure 7.1. Scanning electron micrographs of fetal human ovaries: (a) Note the germinal epithelium is covered with microvilli, blebs of various sizes and solitary cilia (X10,500). (b) Note even distribution of microvilli and blebs on cobblestone-like cells (X12,000).

OVULATORY FOLLICLE

Ovulation is associated with several morphological and cytological parameters, *e.g.*, loosening of granulosa cells, mass, appearance of double-layering of the zona pellucida, meiotic division of the oocyte, polar body formation, loosening of the nuclear envelope, and uniform distribution of the cell organelles.

The ovulation point of the centrally located stigma of the ruptured follicles appears as a crater containing numerous follicular cells, red blood cells, fluids and mucin-like substances. Irregular areas of the apex are ruptured, and the ovum emerges within the corona radiata cells that are obscured by the liquor folliculi and intracellular fluids, which appear as a fine layer of granular material (Figure 7.2). Several biophysical factors facilitate the release of the ovum from the apex of the ruptured follicle, *e.g.*, contractile peristaltic activity of the oviductal musculature and the beat of kinocilia of the fimbriae create currents and countercurrents which facilitate egg pick-up from the follicular cavity.

Granulosa Cells

The growth, maturation and atresia of the oocyte is accompanied and influenced by changes in the granulosa cells. The granulosa cells of the ovarian ooctyes appear polygonal or spindle-like. Most granulosa cells located within the cellular plug result from extrusion of the cumulus oophorus. Granulosa cells not immediately adjacent to the antral cavity are spherical or polyhedral in shape and are embedded in a network of fibrous tissue. The granulosa cells are in close contact and often form cords projecting into the antrum.

The surface of granulosa cells is very irregular due to prominent evagination and/or projections. These cytoplasmic evaginations contain a myosin-like material and may be involved in the movement of cortical granules (Cavalotti *et al.*, 1974). The presence of cytoplasmic evaginations may also be related to the capacity of granulosa cells to bind LH, since the loss of these evaginations corresponds to the loss of LH binding sites (Peluso *et al.*, 1977a,b).

Figure 7.2. Ovary of 30-year-old patient with unexplained infertility. Ovarian biopsy taken by culdoscopy. Note areas denuded of cells (X4000).

Ameboid evaginations of the granulosa cells contain contractile microfilaments which may indirectly contribute to the expulsion of the oocyte. Also, the contractile ameoboid movements of cells of the corona radiata may facilitate the pick-up of the oocyte from the stigma of the follicle to the fimbriae of the oviduct (Motta and Van Blerkom, 1975).

POSTOVULATORY FOLLICLE

Ovulation is associated with loosening of the fibrous structure of the tunica albuginea. The superficial layer of the ruptured membrane curls upwards (Figure 7.3). After ovulation the granulosa cells of the follicles enlarge and the cytoplasm of lutein cells accumulates fine lipid droplets and pigment granules (Crombie *et al.*, 1971).

CONCLUDING REMARKS

The granulosa cells are polyhedral elements which, during the period of follicle growth and antrum formation, contain organelles characteristic of secretory cells. These morphological characteristics indicate that granulosa cells may be involved in the synthesis and secretion of elements of the liquor folliculi, the zona pellucida, and steriod hormone biosynthesis. During the preovulatory phase, some granulosa cells contain numerous lipid droplets, mitochondria with villiform cristae and profiles of the smooth-surfaced endoplasmic reticulum. Major morphological transformations occur in the perifollicular region of the Graafian follicle including the differentiation of the cells of the theca interna into steriodogenic elements. In addition, the presence of smooth muscle cells and cholinergic nerves in the perifollicular stroma present both the potential for neuromuscular junctions and the possibility of neural involvement in ovulation.

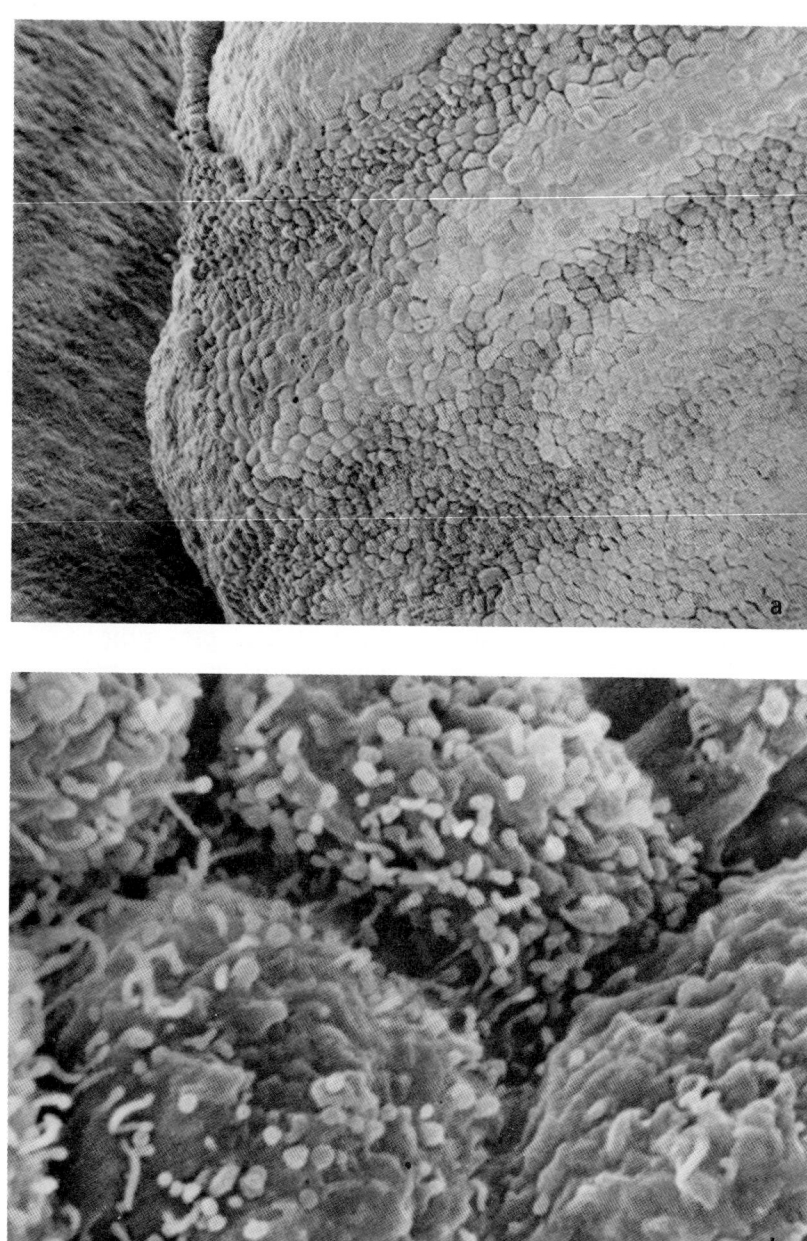

Figure 7.3. Scanning electron micrographs of ovaries from a patient 41 years of age with atrophic ovaries: (a) X800; (b) X20,000.

REFERENCES

Cavalotti, C., G. Familiari, G. Fumagalli, and P. Motta. "Immunofluorescence and Electron Microscopic Observations on the Microfilaments of Ovarian Follicles," *Acta. Histochem.* 52:253 (1974).

Crombie, P. R., R. Burton, and N. Ackland. "The Ultrastructure of the Corpus Luteum of the Guinea Pig," *Z. Zellforsch.* 115:473 (1971).

Hafez, E. S. E., Editor. *Scanning Electron Microscopical Atlas of Mammalian Reproduction* (New York: Springer-Verlag, 1976).

Ludwig, H. and H. Metzger. *The Human Female Reproductive Tract, a Scanning Electron Microscopic Atlas* (New York: Springer-Verlag, 1976).

Motta, P., and J. Van Blerkom. "A Scanning Electron Microscopic Study of the Luteo-Follicular Complex. II. Events leading to Ovulation," *Am. J. Anat.* 143:241 (1975).

Nilsson, O., and S. F. Munshi. "Scanning Electron Microscopy of Mouse Follicles at Ovulation," *J. Submicr. Cytol.* 5:1 (1973).

Peluso, J. J., R. W. Steger, and E. S. E. Hafez. "Sequential Changes Associated with the Atresia of Ovarian Follicles" (1977a).

Peluso, J. J., R. W. Steger, and E. S. E. Hafez. "Surface Ultrastructural Changes in the Granulosa Cells of Atretic Follicles (1977b).

CHAPTER 8

OVIDUCT–UTERUS

M. Oshima, H. Okamura and E. S. E. Hafez

OVIDUCT

The oviductal epithelium is simple columnar, consisting of ciliated cells and nonciliated secretory cells. The proportion of ciliated to nonciliated cells varies, with the greatest number of ciliated cells present in the fimbriae and few in the isthmus. About 70% of the cells in the fimbrial epithelium were ciliated (Figure 8.1). The proportion of ciliated cells decreased gradually from the ampulla to the isthmus, reaching 50% near the ampulla-isthmus junction. Ciliated cells were found singly, or in groups, arranged in rows or in a mosaic pattern (Hafez *et al.*, 1975; Oshima *et al.*, 1975).

In the fimbriae, the ciliated cells are so closely packed that it is impossible to distinguish their boundaries. Nonciliated cells are of uniform diameter (Hafez, *et al.*, 1975) (Figures 8.2 and 8.3).

Cilia first grow at the periphery of the cell and then fill in the central area. Cilia first appear as stubby, short cylinders on those cells that have large numbers of microvilli on their surfaces. Ciliary growth occurs randomly over the mucosal surface, and unevenly over each individual cell. Development of centrioles, which give rise to the necessary number of basal bodies for the induction of ciliary growth, is also asynchronous throughout the epithelium as well as within a single cell (Dirksen, 1971). As cilia increase in length, they are able to bend. There are cells,

108 Scanning Electron Microscopy of Human Reproduction

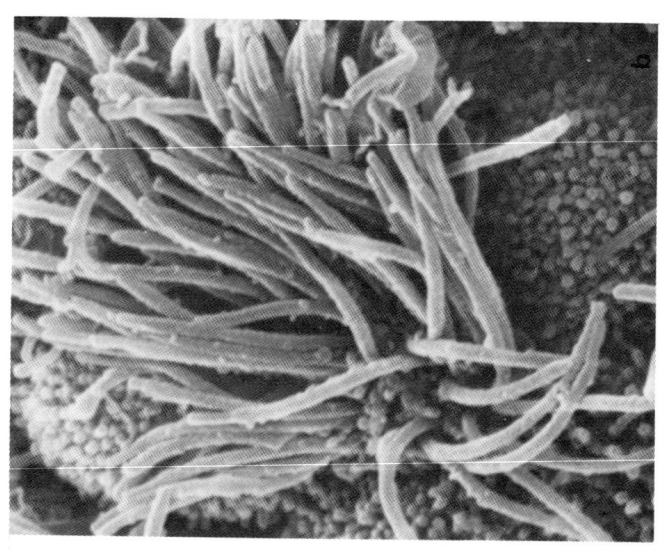

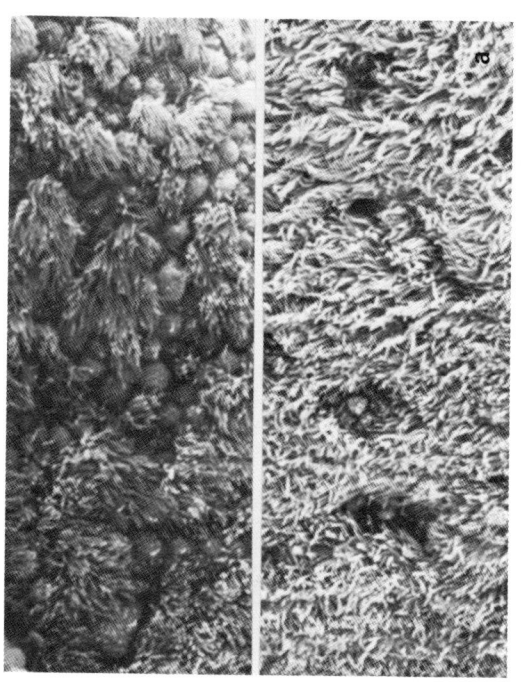

Figure 8.1. (a) Closely packed ciliated cells in human fimbriae (X1000). (b) Cilia in the fimbriae at ovulatory phase (X10,000).

Gynecology 109

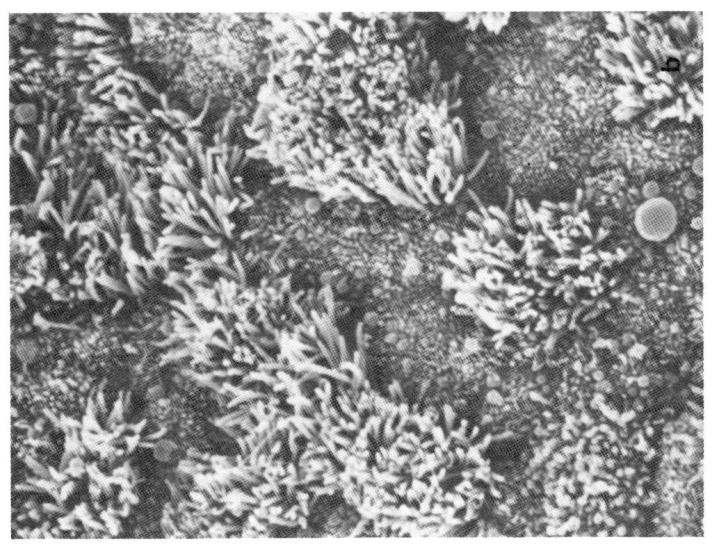

Figure 8.2. (a) Ciliated cells in the ampulla at ovulatory phase (X5000). (b) Ciliated cells in the isthmus at proliferative phase (X5000).

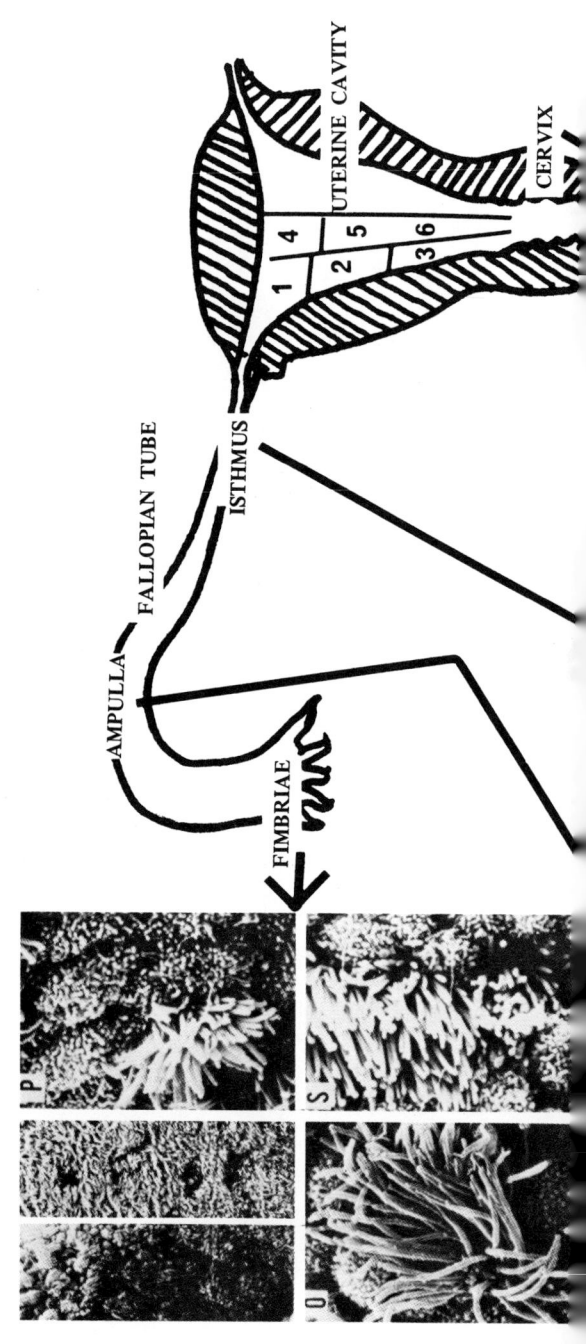

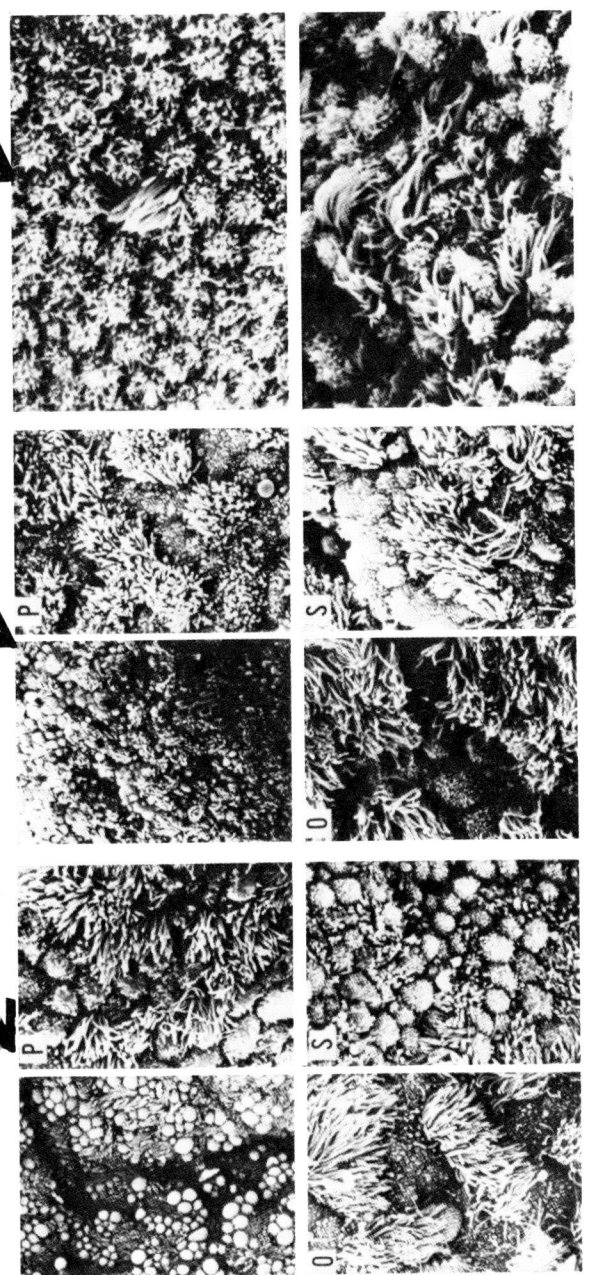

Figure 8.3. Regional differences in the topography and surface ultrastructure of the human oviduct and uterus (right). Note differences between proliferative (p), ovulatory (o), and secretory (s) phases (left).

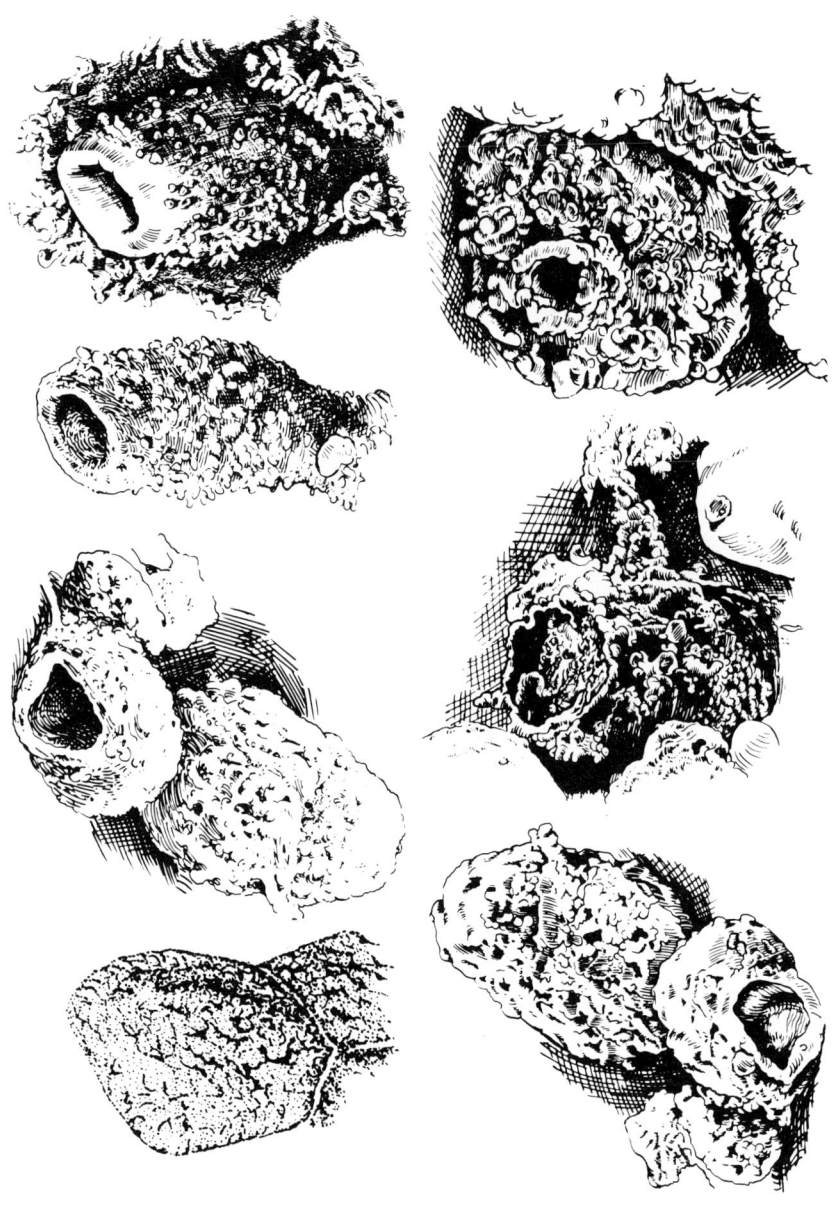

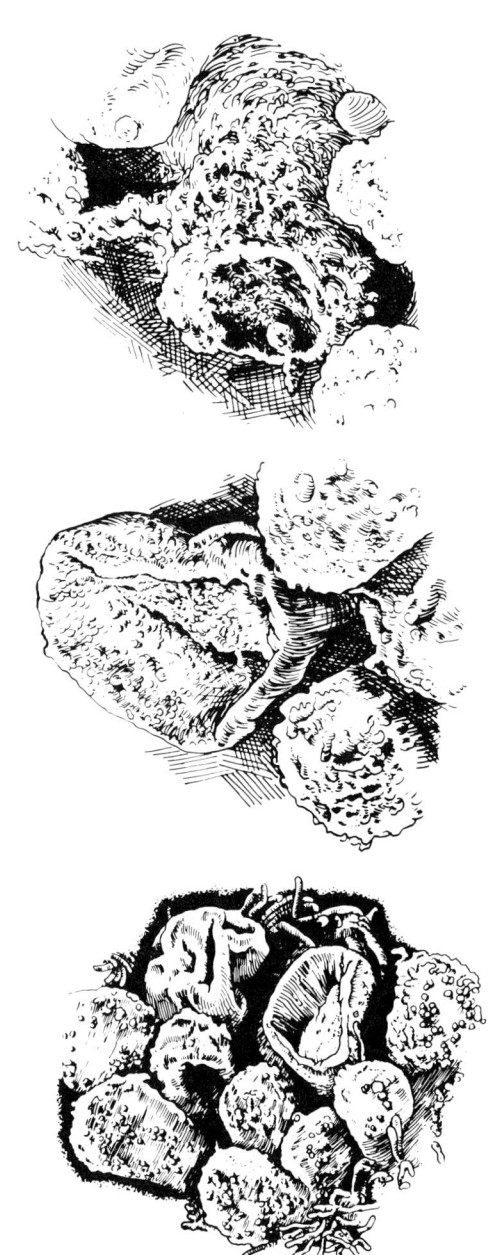

Figure 8.4. Diagrammatic illustration of the nonciliated cells of the human endometrium during different stages of secretory activity.

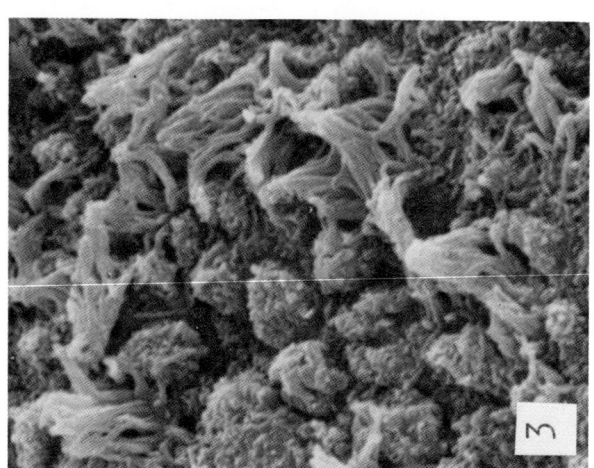
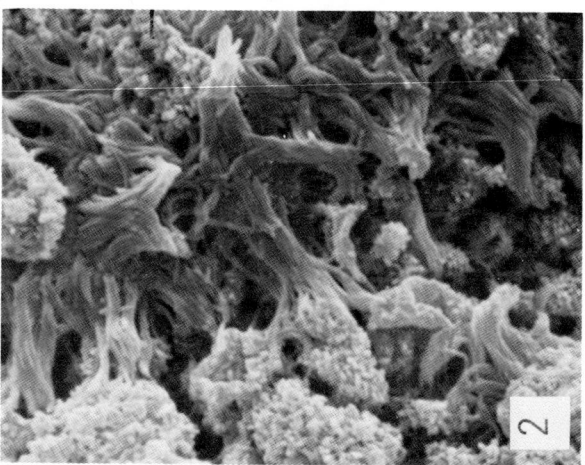
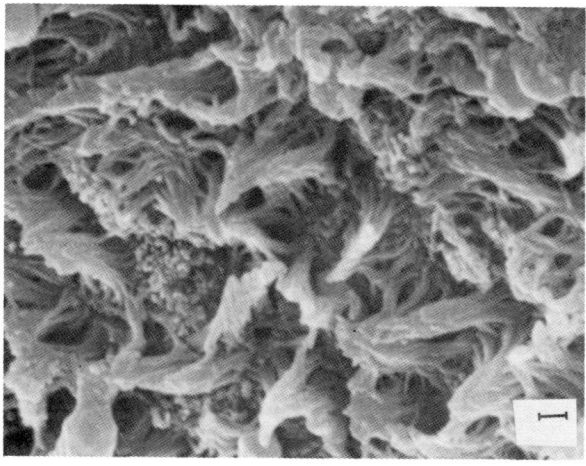

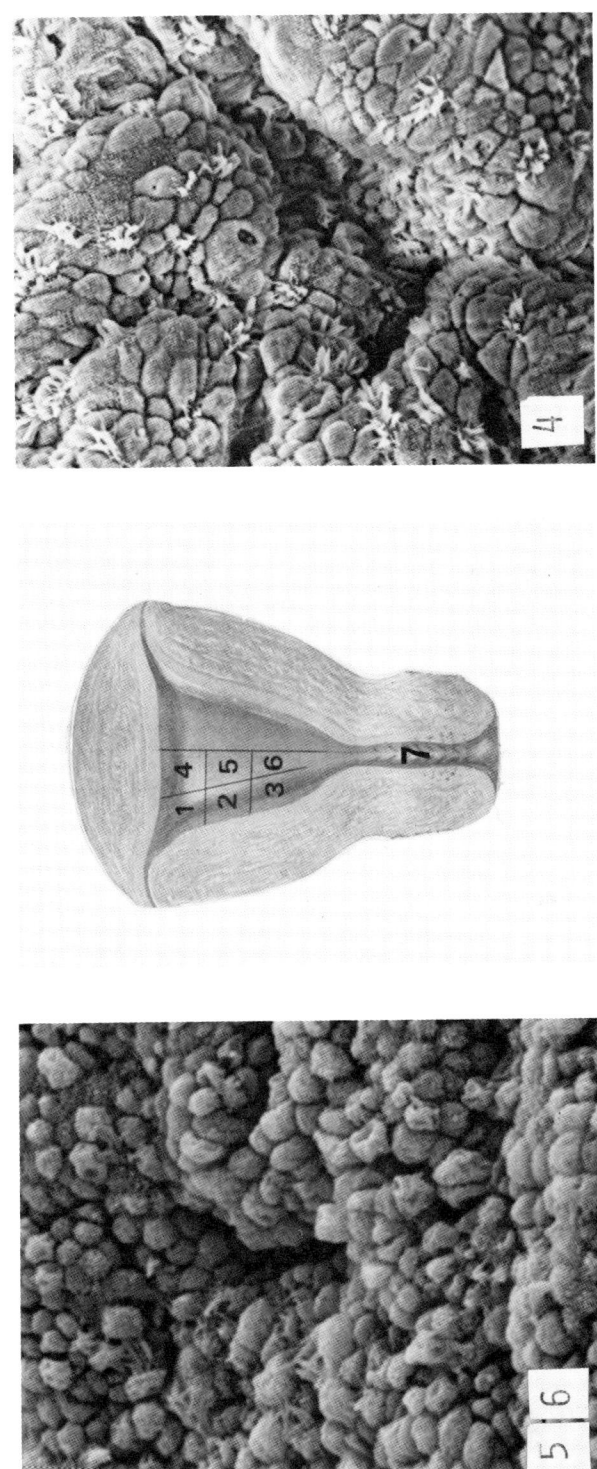

Figure 8.5. Regional differences in surface ultrastructural characteristics of the endometrial cells (X1000-X5000).

however, which at a given age are so undifferentiated that it is impossible to predict their final outcome.

ENDOMETRIUM

The endometrium is smooth with no mucosal folds and the openings of the endometrial glands are apparent. Ciliated cells are less abundant in the endometrium than in the tubal epithelium. They are found singly or in clusters (Figure 8.4). Large cytoplasmic projections are noted at the apical membrane of the nonciliated cells. The abundance, length, shape and interbranching of apical microvilli vary throughout the cycles. Development of apical microvilli, synthesis, storage, release of endometrial secretory granules and ciliogenesis are hormone-dependent. Degenerated cells have been found at random at different stages of the menstrual cycle. Following the release of secretory material from the protruding cell membrane, the cell collapses and becomes wrinkled and devoid of microvilli.

There are regional differences in surface ultrastructural characteristics of the endometrial cells. At the endocervix, very few ciliated cells are arranged in an irregular pattern. Their cyclic changes are not obvious. The density and distribution pattern of ciliated cells vary in various segments of the endometrium. The sparse distribution of ciliated cells on the central part of the uterine wall seems to offer a favorable site for implantation of the blastocyst. On the other hand, ciliated cells appear in clusters in dense string-like formation on the side wall of the uterine cavity (Figure 8.7).

The utero-cervical junction is well recognized by the abundance of ciliated cells in the cervical portion (Figure 8.8). Cilia may play a major physiological role in mucociliary clearance of cervical mucus and in sperm transport to the uterine cavity.

The surface ultrastructure undergoes remarkable changes in various pathological conditions of the endometrium (Oshima, 1974), *e.g.*, endometrial hyperplasia, submucous myoma, endometrial polyp and endometrial cancer (Table 8.1). During postmenopausal life the cells become small and uniform, with small surface protrusions, covered with a few short microvilli. The

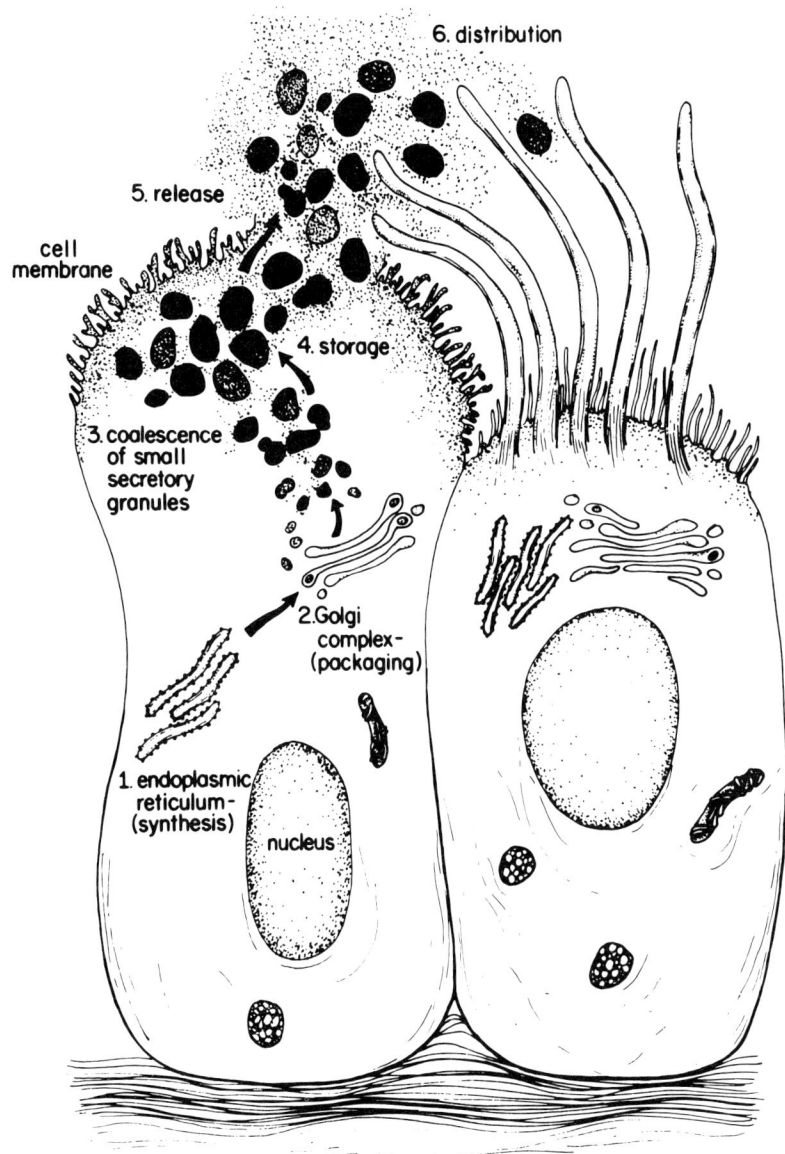

Figure 8.6. Hypothetical representation of one model showing the kinetics of human endometrial fluid. Secretion through the rupture of the apical cell membrane and the release of secretory material (Hafez *et al.*, 1975b). (Courtesy of *Am. J. Obstet. Gynecol.*)

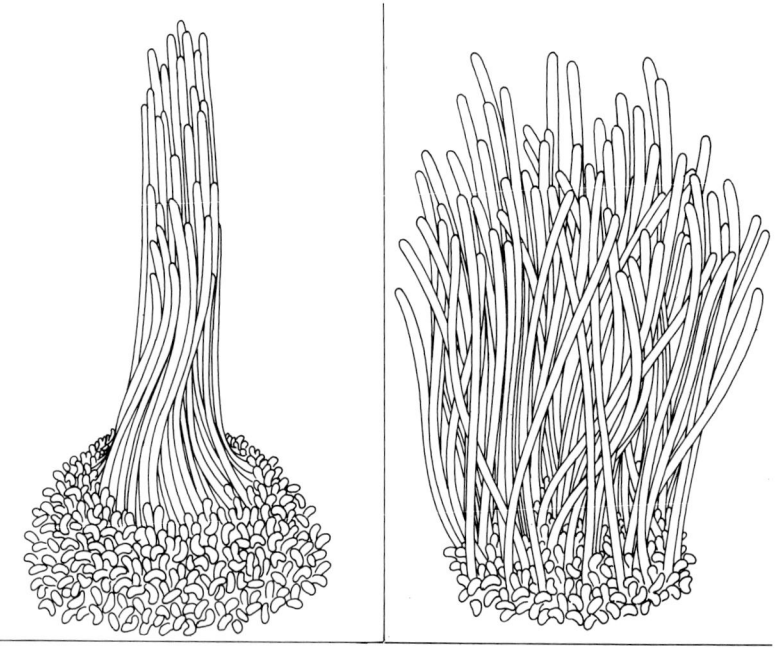

Figure 8.7. Diagrammatic illustration of kinociliated cells in endosalpinx (right) and endometrium (left).

epithelial cells covering polyps show quite wide and flat, free surfaces, forming into polygonal shapes due to swollen intercellular spaces. Microvilli on these cells are short and similar to pinheads.

The epithelial cells overlying the submucous myoma knot vary depending on the degree of extrusion of these knots into the uterine cavity. The endometrium covering these knots shows several atrophic changes as well as overdistention of the cell surface.

Surfaces of cancer cells, quite large in comparison to those found in hyperplasia, are distended into spindle-, semioval- or cylinder-like shapes. Sparse, spot-like microvilli on free surfaces may be interpreted as a sign of undifferentiation or anaplasia of cancer cell surfaces. The morphology of openings of the endometrial glands seems to be characteristic for each cancer type and could be used to differentiate diagnosis.

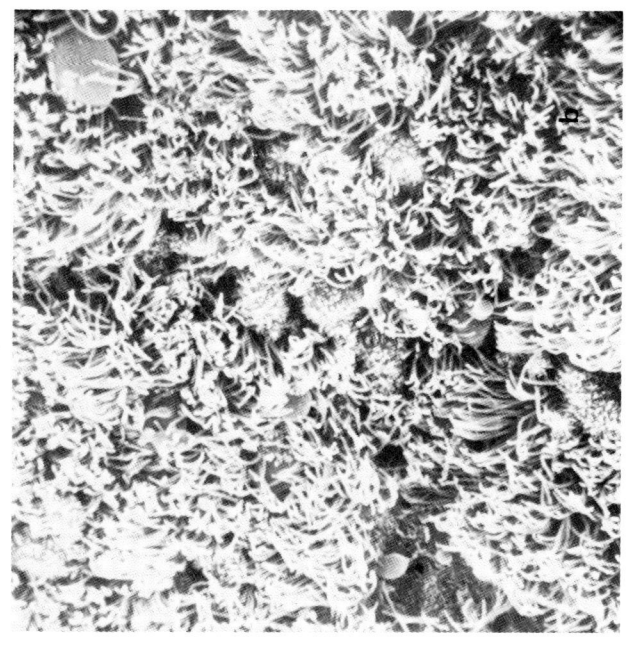
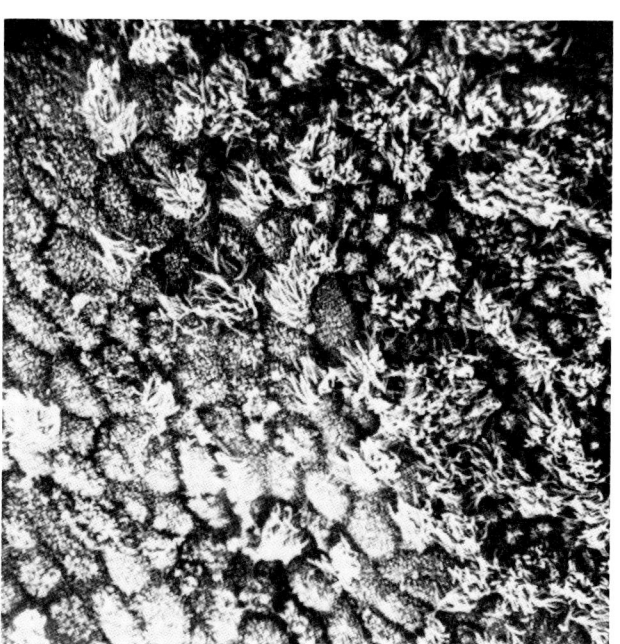

Figure 8.8. Uterocervical junction: (a) Note abundance of ciliated cells in the cervical portion (bottom) and higher frequency of secretory cells in the endometrial portion (top) (X1100). (b) Ciliated cells in the cervical portion just below the uterocervical junction (X2200).

Table 8.1

Surface Ultrastructure Characteristics of Various Pathological Cell Types in the Human Endometrium

Cell Types	Cell Shape	Surface Structure
Atrophic Cells	Small and uniform cells (few protrusions)	Fine and uniform microvilli
Inflammatory Cells	Large uniform cells, ovoid or cylindrical in shape	Uniform distribution of microvilli
Necrotic Cells	Irregular contours	Sparse, irregularly distributed microvilli—sometimes completely absent
Endometrial Hyperplasia	Cell size mostly large, but not uniform	Similar to mid-proliferative phase. Small gland opening, dense microvilli, but absence of spherical protrusions and hypertrophied microvilli
Endometrial Polyps	Cell size variable, sometimes polygonal in shape	Variable, sometimes exhibit swollen intercellular parts with microvilli 0.1μ in length and width
Submucosa Myomata	Varying degrees of cell surface distension. Cells vary in size	Variable, depending on degree of protrusion. Variable degree of atrophy
Endometrial Cancer	Cells are usually large and often distended into spindle-, semioval- or cylinder-like shapes	Cell surfaces often rough and irregular. Protrude from necrotic tissue. Sparse microvilli

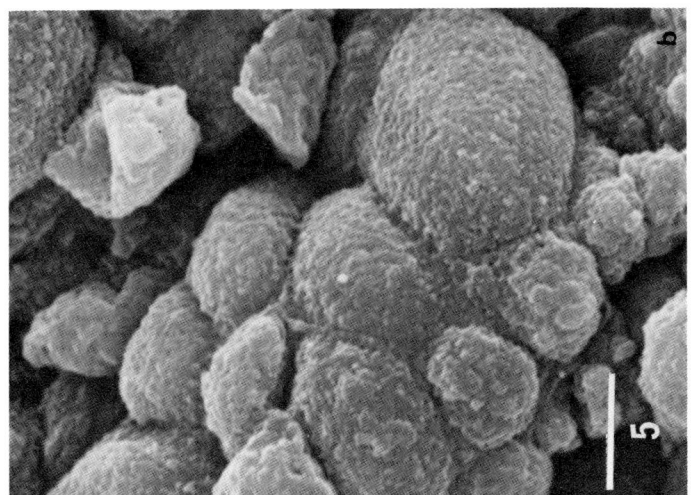
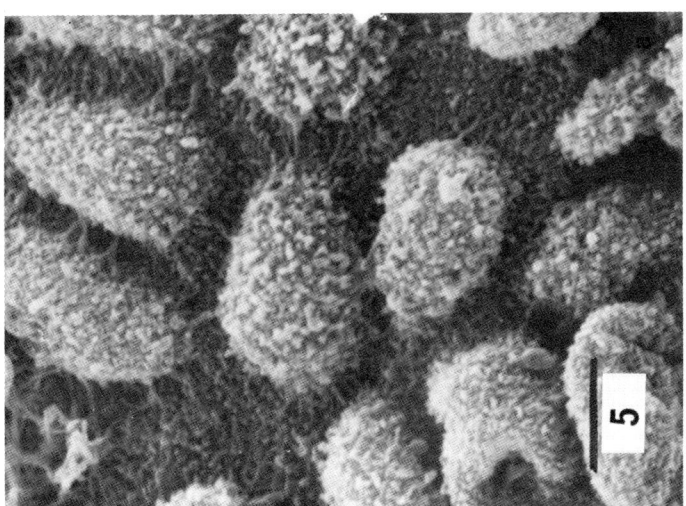

Figure 8.9. Various pathological changes of human endometrium: (a) Menopausal atrophic changes (X3000). (b) Hyperplasia (X3000).

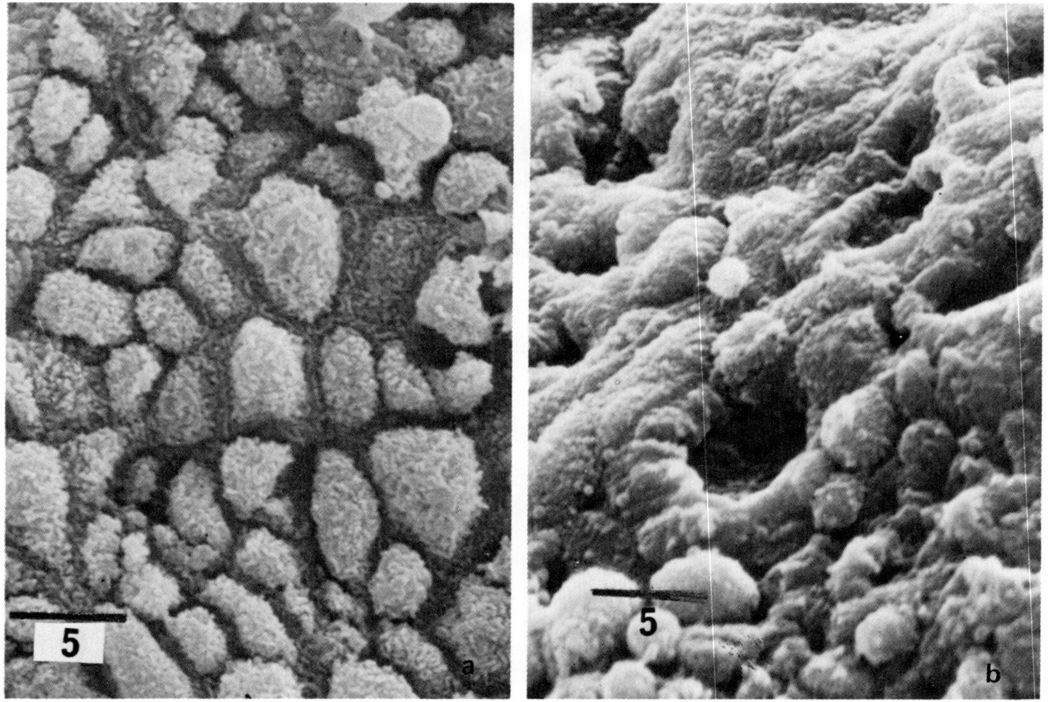

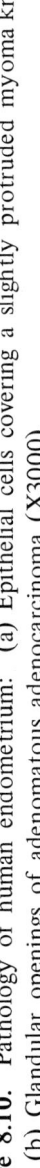

Figure 8.10. Pathology of human endometrium: (a) Epithelial cells covering a slightly protruded myoma knot (X3000). (b) Glandular openings of adenomatous adenocarcinoma (X3000).

Gynecology 123

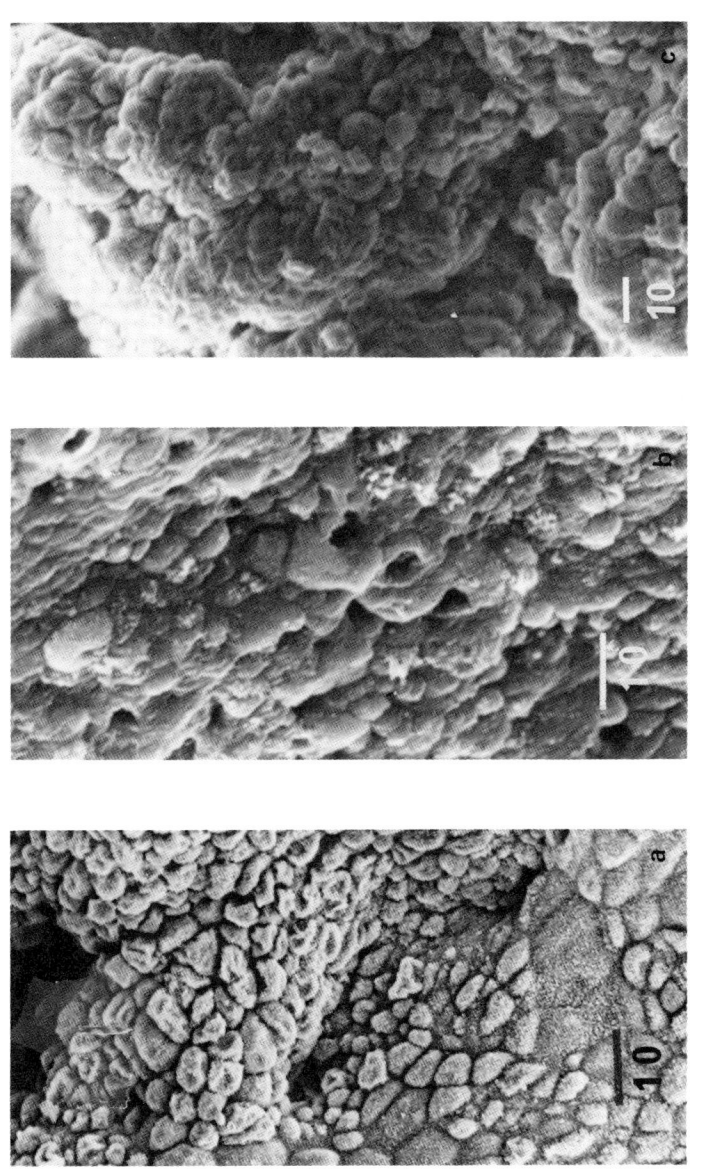

Figure 8.11. Pathology of human endometrium: (a) Tubular adenocarcinoma of the human endometrium (X1000). (b) Adenomatous adenocarcinoma (X1000). (c) Papillary adenocarcinoma (X500).

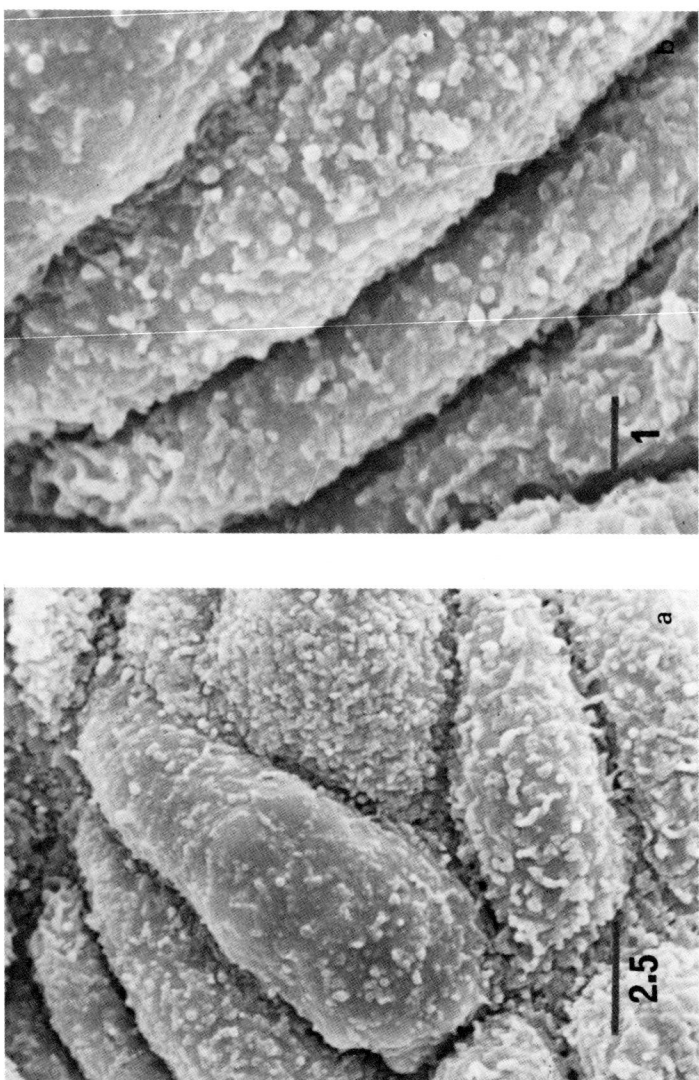

Figure 8.12. Pathology of human endometrium: (a) Higher magnification of endometrial cancer cells (X5000). (b) Microvilli on endometrial cancer cells are sparse and spot-like (X10,000).

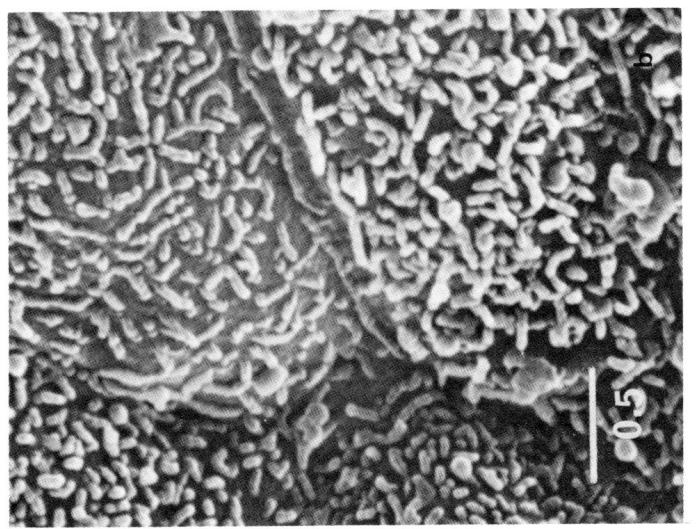

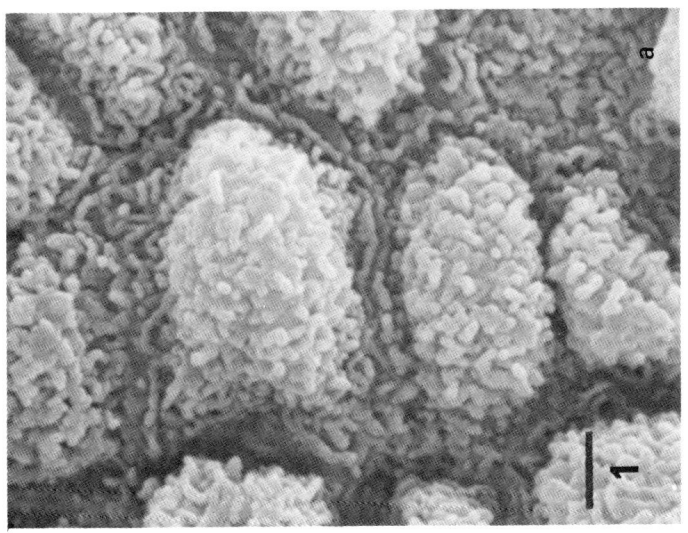

Figure 8.13. Pathology of human endometrium: (a) Slightly protruded submucous myoma (X10,000). (b) Semi-spherically protruded submucous myoma (X20,000).

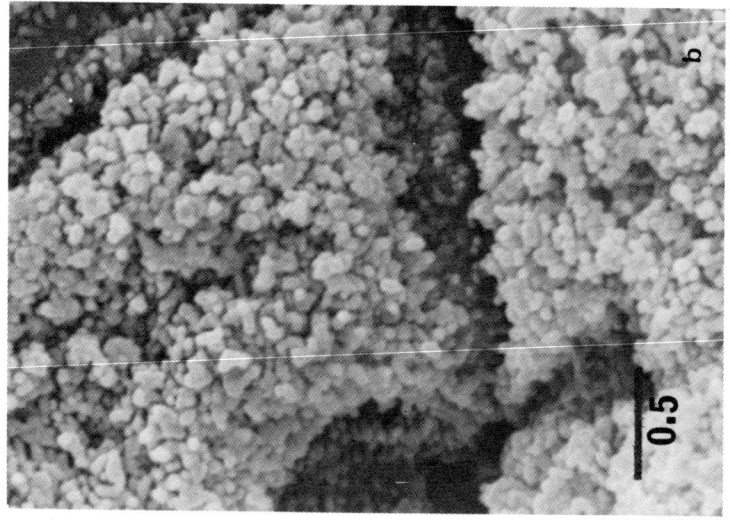
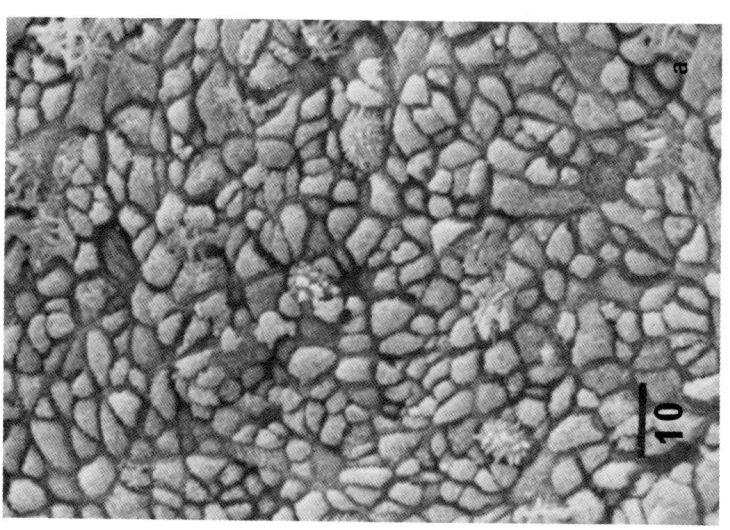

Figure 8.14. Pathology of human endometrium: (a) Atrophic findings on polyp-like submucous myoma (X1000). (b) Microvilli on polyp-like submucous myoma (X20,000).

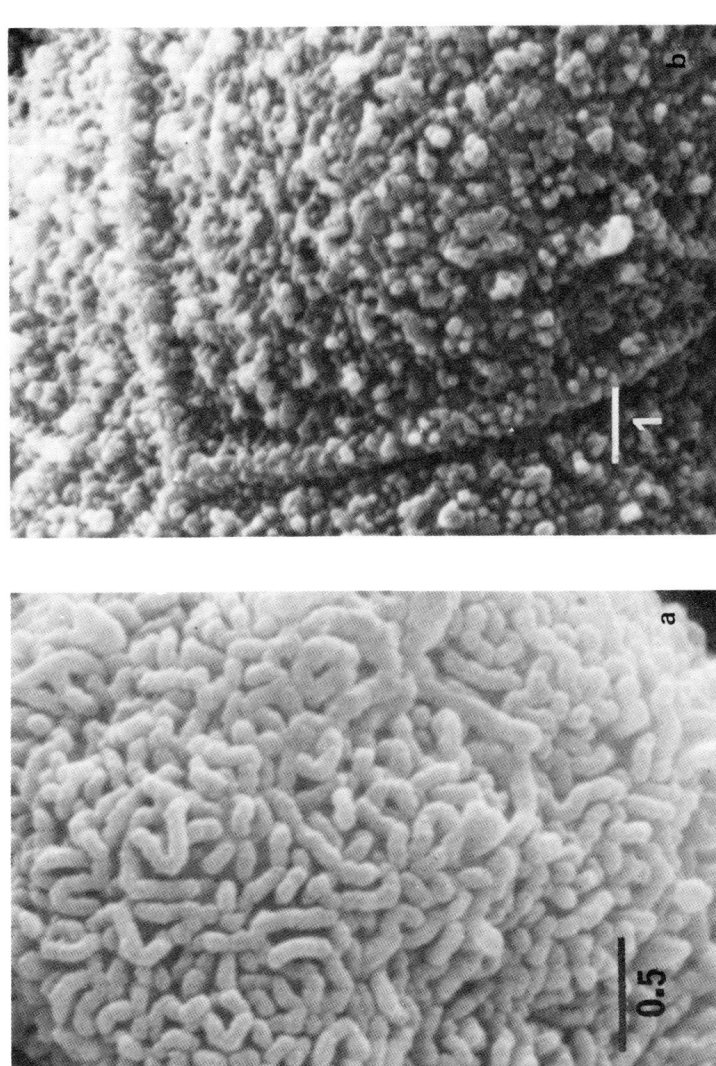

Figure 8.15. Pathology of human endometrium: (a) Endometrial hyperplasia (X20,000). (b) Endometrial polyp (X10,000).

FUNCTIONAL SIGNIFICANCE

Extensive studies have been conducted on the scanning electron microscopical structure of the human oviduct (Bonilla-Musoles et al., 1973; Ferenczy et al., 1972), endometrium (Bank et al., 1975; Ferenczy et al., 1973, 1976; Hafez, 1972; Hafez et al., 1975a,b; Johannisson and Nilsson, 1972; Ludwig and Metzger, 1976; Nilsson and Hagenfeldt, 1973) and the cervix (Williams et al., 1973; Zaneveld et al., 1975). The epithelial cells of the female reproductive tract have several mechanisms of transport processes across the plasma membrane. These processes regulate cellular volume, nutrition, excretion and communication along the surface of the cell as well as the intracellular and extracellular communication. The luminal fluids play a major role in the capacitation of spermatozoa, and the nutrition of the embryo. The oviductal and uterine fluids are made of (1) components from the transudation of blood serum; and (2) protein, carbohydrate and other metabolites synthesized within the epithelial cells and discharged through the apical cell membrane. The synthesized polypeptides are transported to the Golgi complex where the immature secretory granules undergo maturational changes and coalesce before they are discharged into the lumen. The release of secretory material from the epithelial cells seems to occur over several days during the menstrual cycle. This is achieved by asynchronous discharge of adjacent cells in the same region.

Degenerated epithelial cells may or may not be secretory cells which have already discharged their contents. It is not known whether each cell discharges secretions only once or more than once during its lifetime. The secretory material is not simply a nonspecific deposit of luminal secretion product, for it is absent from the surface of cilia. The physiological mechanisms that control the synthesis of luminal fluids are different from those that control discharge of secretions into the lumen of the female reproductive tract.

The endometrium selectively retains necessary metabolites during the process of filtration, and regulates the composition of endometrial fluid through facultative absorptive and secretory processes (Keynes, 1969). This active transport applies to both secretory and absorptive functions depending on the net direction of movement of the substrate molecules to and from the cell across the cell membrane via the carrier protein.

During the menstrual cycle, the cyclical differentiation and regression of cytoplasmic organelles may be associated with self-digestion by lysosomes (autolysosomes). Lysosomes appear in higher concentrations in endometrial cells during the late follicular phase of the menstrual cycle. They contain cellular debris in several stages of digestion, some arising from the ingestion of adjacent epithelial cells, *e.g.*, leukocytes, and some from portions of the cell's own cytoplasm, which have been released in the lumen (Lawn, 1973).

The membrane of epithelial cells is hormonally controlled, as judged by the cyclical changes in distribution, shape and number of apical microvilli, as seen in the secretory cells throughout the menstrual cycle. In the female reproductive tract, kinocilia play an important role in the transport of particles and in directing the flow of luminal fluids.

The resolving power of the electron microscope theoretically would have allowed one to visualize the arrangement of the mucus constituents, provided that one could preserve the arrangement of glycoproteins in a gel consisting mostly of water.

Light microscopy and transmission electron microscopy have been of little use in the study of the ultrastructure of cervical mucus due to the marked scattering of the electron beam through samples of sufficient thickness to give a three-dimensional view of the gel structure. Transmission electron microscopy has neither permitted one to observe the spatial distribution of the gel macromolecules nor allowed one to detect clear differences in gel structure during the ovarian cycle. Singer and Reid (1970) described in human cervical mucus the existence of fibers organized as parallel filaments. Elstein *et al.* (1971), on the other hand, observed this material to be composed of globular structures arranged in chains, whereas Van Bruggen and Kremer (1970) have described the presence of filaments of varied dimensions in bovine and human mucus. The validity of the results reported in the two later papers is questioned since the native structure of the gel was altered either by dilution or by homogenization, prior to observation.

REFERENCES

Bank, H. L., H. O. Williamson and K. Manning. "Scanning Electron Microscopy of Copper-Containing Intrauterine Devices: Long-Term Changes *In Utero*," *Fertil. Steril.* 26:503-512 (1975).

Bonilla-Musoles, F., J. Hernandez-Yago and J. Renau. "Microscopia Electronica de rastero del epitelio de la trompa de Falopio," *Rev. Esp. Obstet. Ginecol.* 32:305-311 (1973).

Dirksen, E. R. "Centriole Morphogenesis in Developing Ciliated Epithelium of the Mouse Oviduct," *J. Cell Biol.* 51:286 (1971).

Elstein, M., R. F. Mitchell and J. T. Syrett. "Ultrastructure of Cervical Mucus," *J. Obstet. Gynaecol. Brit. Commonwlth.* 78:180 (1971).

Ferenczy, A., R. M. Richart, F. J. Agate, Jr., M. L. Purkerson and E. W. Dempsey. "Scanning Electron Microscopy of the Human Fallopian Tube," *Science* 175:783-784 (1972).

Ferenczy, A. and R. M. Richart. "Scanning and Transmission Electron Microscopy of the Human Endometrial Surface Epithelium," *J. Clin. Endocrinol. Metab.* 36:999-1008 (1973).

Ferenczy, A. "Studies on the Cytodynamics of Human Endometrial Regeneration. I. Scanning Electron Microscopy," *Am. J. Obstet. Gynecol.* 124:64-74 (1976).

Hafez, E. S. E. "Scanning Electron Microscopy of the Female Reproductive Tract," *J. Reprod. Med.* 9:119-134 (1972).

Hafez, E. S. E. *Scanning Electron Microscopy Atlas of Mammalian Reproduction* (New York: Springer, 1975).

Hafez, E. S. E. and H. Kanagawa. "Endometrial-Blatocyst Interaction (Rabbit)," in *Scanning Electron Microscopic Atlas of Mammalian Reproduction*, E. S. E. Hafez, Ed., (New York: Springer, 1975).

Hafez, E. S. E., M. I. Barnhart, H. Ludwig, J. Lusher, I. Joelsson, J. L. Daniel, A. I. Sherman, J. A. Jordan, H. Wolf, W. C. Stewart and F. C. Chrétien. "Scanning Electron Microscopy of Human Reproductive Physiology," *Acta Obstet. Gynecol. Scand.* Suppl. 40 (1975a).

Hafez, E. S. E., H. Ludwig and H. Metzger. "Human Endometrial Fluid Kinetics as Observed by Scanning Electron Microscopy," *Am. J. Obstet. Gynecol.* 122:929-938 (1975b).

Hafez, E. S. E. and H. Ludwig. "Scanning Electron Microscopy of the Endometrium," in *Biology of the Uterus*, R. M. Wynn, Ed. (New York: Plenum Press, 1977).

Johannisson, E. and L. Nilsson. "Scanning Electron Microscopic Study of the Human Endometrium," *Fertil. Steril.* 23:613-625 (1972).

Keynes, R. D. "From Frog Skin to Sheep Rumen: a Survey of Transport of Slats and Water across Multicellular Structures," *Quart. Rev. Biophys.* 2:177-281 (1969).

Lawn, A. M. "The Ultrastructure of the Endometrium during Sexual Cycle," in *Advances in Reproduction Physiology*, M. W. H. Bishop, Ed. (London: Elek Science, 1973), Chapt. 2.

Ludwig, H. and H. Metzger. *The Human Female Reproductive Tract. A Scanning Electron Microscopic Atlas* (Berlin: Springer-Verlag, 1976).

Ludwig, H. and H. Metzger. "Zur Ultrastruktur der Tubeninnenflache im Raterelektronenmikroskop," *Arch. Gynaekol.* 210:251 (1971).

Masterson, R., E. M. Armstrong and I. A. R. More. "The Cyclical Variation in the Percentage of Ciliated Cells in the Normal Human Endometrium," *J. Reprod. Fertil.* 42:537-540 (1975).

Nilsson, O. and K. Hagenfeldt. "Scanning Electron Microscopy of Human Uterine Epithelium Influenced by the TCU Intrauterine Contraceptive Device," *Am. J. Obstet. Gynecol.* 117:469-472 (1973).

Oshima, M. "A Scanning Electronmicroscopic Study on the Human Endometrium in Various Pathologic Conditions," *Adv. Obstet. Gynecol.* 26: 209 (1974).

Oshima, M. O. Harada, H. Okamura and T. Nishimura. "The Distribution and Cyclic Changes of Ciliated Cells in the Female Reproductive Tract," *J. Clin. Elec. Micros.* 8:451 (1975).

Singer, A. and B. L. Reid. "The Ultrastructure of Cervical Mucus," *J. Reprod. Fertil.* 21:377 (1970).

Van Bruggen, E. F. J. and J. Kremer. "Electron Microscopy of Bovine and Human Cervical Mucus," *Internat. J. Fertil.* 15:50 (1970).

Williams, A. E., J. A. Jordan, J. M. Allen and J. F. Murphy. "The Surface Ultrastructure of Normal and Metaplastic Cervical Epithelia and of Carcinoma *In Situ*," *Cancer Res.* 33:504-513 (1973).

Zaneveld, L. J. D., P. F. Tauber, C. Port, D. Propping and G. F. B. Schumacher. "Structural Aspects of Human Cervical Mucus," *Am. J. Obstet. Gynecol.* 122:650-654 (1975).

CHAPTER 9

UTEROTUBAL JUNCTION*

Hossam E. Fadel

The way in which the fallopian tube joins the uterine cavity differs in various mammalian species (Hafez and Black, 1969). In the human, as well as in some other species, *e.g.*, the rhesus monkey, a well-defined intramural portion of the tube courses through the uterine wall before opening into the cornual angle of the endometrial cavity. This portion of the tube and the adjoining portion of the uterine cornu serve many physiological roles in mammalian reproduction. It is no wonder that this area receives much attention in the clinical evaluation of infertility problems. Recently, interest in this area has intensified because of the possibility of achieving female sterilization by its occlusion or interference with its proper functioning through a transcervical approach (Brueschke *et al.*, 1977). The interstitial portion of the tube is 1.0-2.5 cm long and narrow; its lumen measures 100-400 µm. It describes quite a variable course through the uterine wall. Sweeney (1962) reported that 8% of the tubes followed a gentle curve, 23% a straight course, while the remainder followed a tortuous course either angulated or with 2-4 convolutions.

Many investigators have studied the musculature of the intramural portion. However, dispute over the presence of a

---

*This investigation was supported by the Contraceptive Development Branch, Center for Population Research, National Institute of Child Health and Human Development, National Institutes of Health, under Contract NIH-NICHD-71-2233.

sphincteric mechanism continues (David and Czernobilsky, 1968; Rocker, 1964). On the other hand, few have studied the epithelial lining of this portion (Lisa et al., 1954; Fadel et al., in preparation). More recently, the scanning electron microscope (SEM) has provided an opportunity to study the morphological details of surface epithelia. Utilizing the SEM and using a special dissection technique we have been able to study the surface structure of this portion of the tube and its merger with the cornual endometrium (Fadel et al., 1976a,b). This area (Figure 9.1a), which has been collectively referred to as the uterotubal junction (UTJ) will be described here. (The surface epithelium of the remainder of the tube (extramural portion) and endometrium is described elsewhere in this book).

SCANNING ELECTRON MICROSCOPIC CHARACTERISTICS OF UTJ

Both the interstitial endosalpinx and endometrium consist of ciliated and secretory (nonciliated) cells. They have many similar morphologic and functional characteristics (Figure 9.1b). They respond to ovarian hormonal influences in a similar way. Both show cyclic changes that are comparable, but quantitatively much less than the cyclic changes occuring in the uterine glandular epithelium. Because of these changes the SEM appearance of the UTJ will be described separately in the: (1) Proliferative Phase, (2) Secretory Phase, and (3) Postmenopausal Period.

The Proliferative Phase

Ciliated cells comprise 5% and 10% of the cell population in the cornual endometrium and interstitial endosalpinx, respectively. Endometrial glands are present in the cornual endometrium (Figure 9.1c). Ciliated cells are found in greater numbers around and within the mouths of the glands. There are 50-60 cilia per cell. The ciliary shafts are well developed and measure approximately 4-6 μm in length. Secretory cells are dome-shaped. Their surfaces present numerous, rounded regular and evenly distributed microvilli (Figure 9.1b).

The transitional area between the endosalpinx and endometrium is characterized by: (1) a marked increase in the number of ciliated cells—the ratios between ciliated and nonciliated cells become 1:2 or 1:1; and (2) the secretory cells tend to be flattened and assume a polygonal elongated shape. Both changes are increased noticeably in both the endometrial and endosalpingeal cells with proximity to the transitional area (Figure 9.1d).

The Secretory Phase

Towards midcycle, apical cytoplasmic protrusions of some secretory cells of both the endometrium and endosalpinx become apparent. The microvilli on these protrusions are more greatly separated from each other than on the rest of the cell wall. These protrusions, which represent apocrine secretory material, become larger and more numerous in the early secretory phase (Figure 9.2a). Later, in the luteal phase, these protrusions become emptied of secretions and the cell appears to be collapsed with transparent walls (Figure 9.2b).

The secretory activity is more evident in the endometrium than in the endosalpinx. In the latter, a third cell-type is seen, the "peg" cell (Figure 9.2c). This has a characteristic irregular, wrinkled surface with scant microvilli. This cell represents an exhausted and/or degenerated secretory cell. Not all the cells go through these different stages of the secretory process simultaneously. Actually it is more common to find inactive-looking cells, cells with secretory globules of varying size, collapsed cells, and/or ruptured cells alongside each other in the same field, both in the endometrium and endosalpinx (Figure 9.2d). The transitional area showed the same changes and bore the same characteristic features of marked increase in the ratio of ciliated cells and flattening of the secretory cells.

There is some disagreement about cyclic changes in the ciliated cells. We did not notice changes in either the number, distribution or structure of ciliated cells in the secretory phase (Fadel *et al.*, 1976a,b). Previous observations indicated a peak occurrence of ciliated cells in the endometrium at mid-cycle with a decrease during the luteal phase (Ferenczy and Richart,

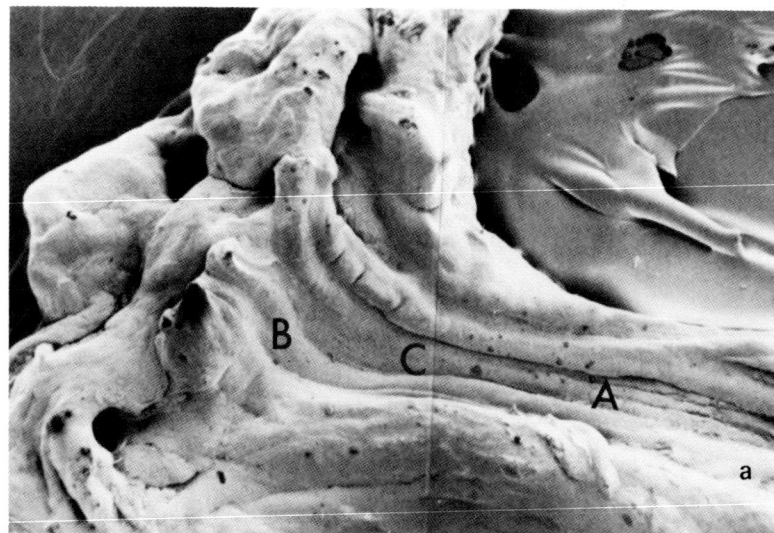

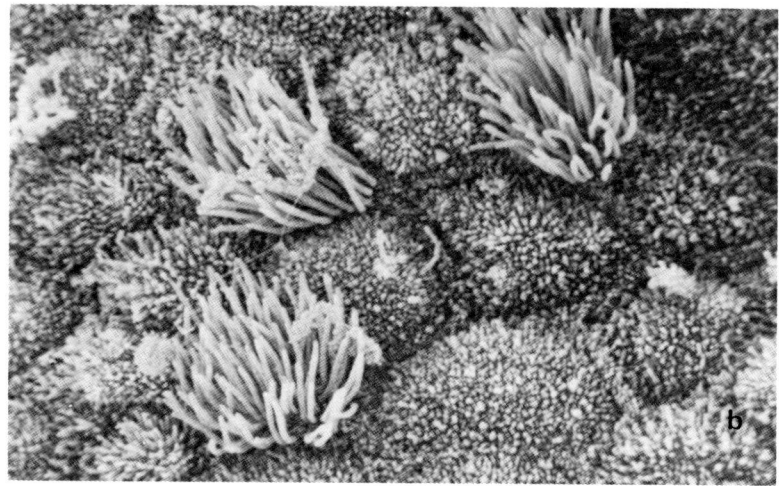

Figure 9.1. (a) The Uterotubal Junction (X12). A low SEM microphotograph showing (A) Interstitial endosalpinx with its characteristic plicated mucosa (3-4 folds); (B) Cornual Endometrium; and (C) Transitional area where the endosalpinx merges with the endometrium. These three areas collectively are referred to as the uterotubal junction (UTJ).
(b) Ciliated and secretory cells (X3000). The Secretory cells are rounded and dome-shaped. Their apical surfaces are evenly covered with multiple, rounded, regular microvilli. There are about 50-60 cilia per cell. The cilia are well-preserved, vigorous, and 4-6 μm in length. This microphotograph represents endometrial cells, but the cells of the endosalpinx are similar to these.

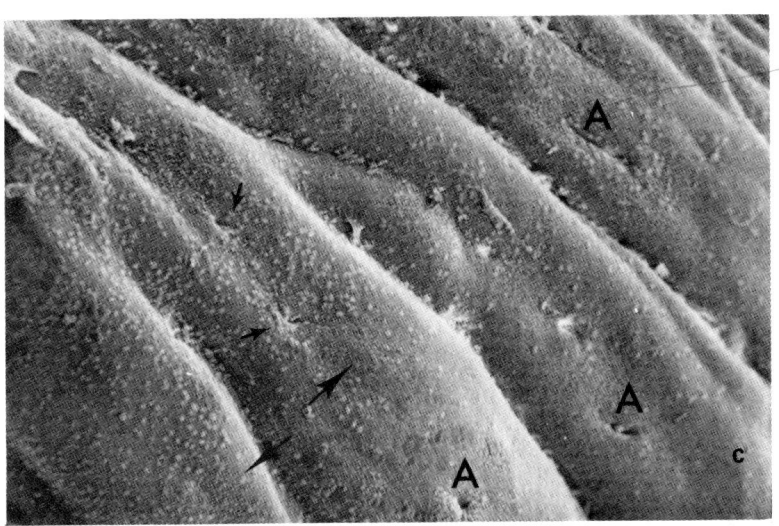

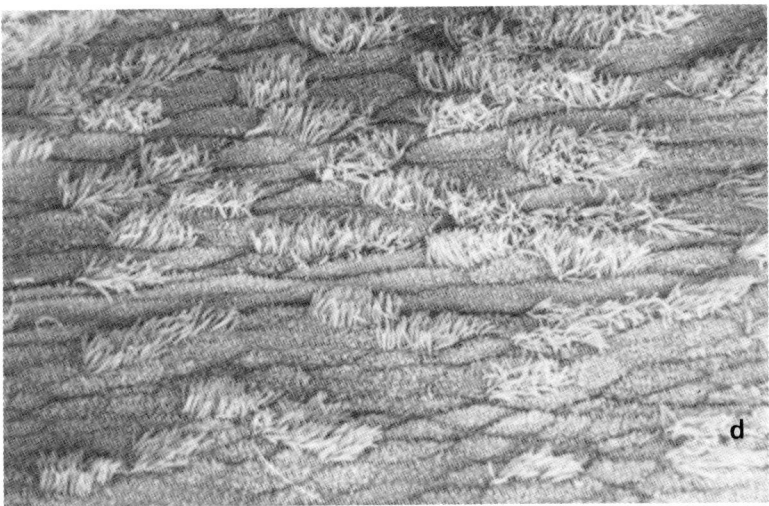

Figure 9.1. (c) Cornual endometrium: proliferative phase (X100). The slightly convoluted endometrial surface is seen. Openings of endometrial glands are shown (A). On closer inspection, tufts of cilia can be distinguished. They are particularly numerous around the mouths of the endometrial glands (small arrows). Secretory cells can also be identified (big arrows). The ratio of ciliated to nonciliated cells is approximately 1:20.
(d) Transitional area (X1000). Note the increased number of ciliated cells. Also, note that the secretory cells are flattened, elongated and polygonal in shape.

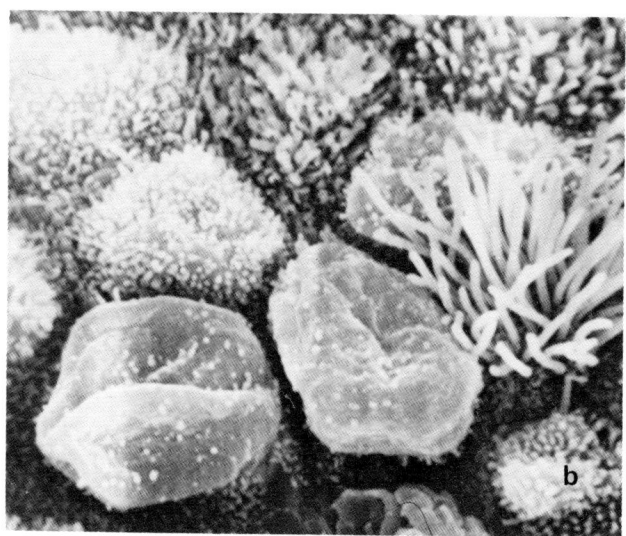

Figure 9.2. (a) Cornual endometrium: early secretory phase (X2000). These are cornual endometrial secretory cells showing apical cytoplasmic (secretory) protrusions of varying sizes (arrows). The same type of activity is seen in endosalpinx, but to a quantitatively lesser extent.
(b) Cornual endometrium: late secretory phase (X5000). Emptied secretory cells. The cell walls are transparent. Note the well-preserved ciliary tufts. There is no blistering, drooping, or other degenerative change in the cilia. [The micrographs in this figure are reproduced by permission from *Fertil. Steril.* 27:1176 (1976).]

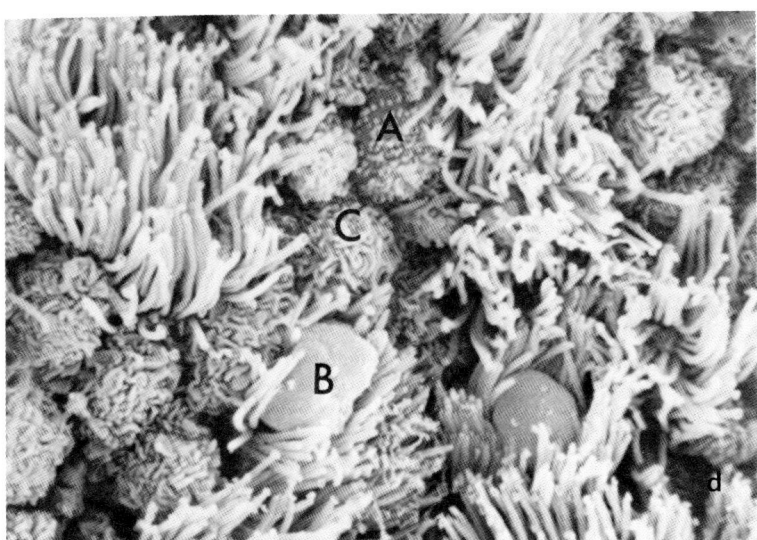

Figure 9.2. (c) Interstitial endosalpinx: secretory phase—"peg" cells (X3000). These are cells with wrinkled walls and sparse microvilli. Note that the ciliated cells are well-preserved and the cilia are healthy and vigorous-looking.
(d) Interstitial endosalpinx: secretory phase (X5000). Secretory cells in different stages of activity. Some are seen with secretory protrusions (A); others are exhausted/degenerated "peg" cells (B); still others have prominent microvilli and microridges (C). [The micrographs in this figure are reproduced by permission from *Fertil. Steril.* 27:1176 (1976).]

1973). Clumping, drooping and blistering of the cilia of both the endometrium (Johannisson and Nilsson, 1972) and endosalpinx (Patek, 1974) have been described. These differences may be due to different hormonal sensitivity of the cells in the UTJ, which were not studied by these investigators, and the cells in the other parts of the endometrium and endosalpinx. Also, the structural changes (blistering, drooping) may be artefactual due to the use of freeze-drying in the preparation of specimens. We used critical point drying, the latter method having been credited with better preservation of fine details and delicate features of the cells.

Postmenopausal Period

In women 2-6 years after the menopause, we found that the secretory cells of both the cornual endometrium and interstitial endosalpinx appear lower and more flattened, but there was no atrophy, degeneration, or desquamation. The cells were polygonal rather than rounded and appeared to be separated further from each other, particularly in the endosalpinx. There was no evidence of secretory activity. The openings of the endometrial glands were wider and appeared more prominent, probably due to flattening of the surrounding epithelium. The ciliated cells were present in the usual ratio to the secretory cells. The cilia were well developed, vigorous and healthy looking (Figure 9.3a,b,c). In the transitional area, the flattening of the cells was quite marked. The intercellular spaces were wider and more distinct in this area than in the adjoining portions of the endometrium and endosalpinx. There was also a relative increase in the number of ciliated cells, but not as marked as in the specimens obtained from women in the childbearing period (Figure 9.3d).

Different findings were reported by others. Marked decrease in the number of ciliated cells was observed in the isthmic, but not in the infundibular or ampullary portions of the tube, and only in women who were more than 15 years past the menopause (Patek et al., 1972; Patek, 1974). In yet another study it was reported that the ciliary shafts were rarely encountered in the tubal mucosa of postmenopausal women

(Ferenczy and Richart, 1974). In another study of the tubal epithelium, there was a marked decrease in the number of ciliated cells, particularly in the infundibulum. There were fewer cilia per cell and they were shorter. Also noted was apparent sloughing off of epithelial cells, both secretory and ciliated, in localized areas, particularly in older patients (Gaddum-Rosse et al., 1975). In an SEM study of endometrium recovered from advanced postmenopausal women, the cells exhibited a uniform cobblestone pattern, were covered by short sparse microvilli, and were devoid of ciliated cells or secretory activity (Ferenczy and Richart, 1974). Again it must be remembered that these studies included neither the interstitial portion of the tube nor the cornual endometrium immediately adjacent to UTJ, which may explain the disagreement with our results. It may be that estrogens are important in both ciliogenesis and maintenance of the ciliated cells in other parts of endometrium (Scheuller, 1973) and the fallopian tube (Gaddum-Rosse et al., 1975). The latter investigators had shown that long-, but not short-term estrogen therapy in postmenopausal women resulted in normal ciliation of the infundibular epithelium, and more effective ciliary activity. Animal experiments indicate that in several species (e.g., rabbits, rhesus monkeys), ciliogenesis and maintenance of the ciliated cells are clearly dependent on estrogen (Brenner and Anderson, 1973; Rumery and Eddy, 1974), while progesterone induced deciliaton and dedifferentiation in the oviductal epithelium of the rhesus monkey (Brenner and Anderson, 1973). The effect of progesterone administration to women has also been studied. In fertile women, low-dose gestagen causes the epithelium of the fallopian tube to have a similar appearance to that in the late secretory phase, while long-acting progestin given to a woman 25 years after the menopause caused extensive deciliation in the tube (Patek, 1974).

Physiological Correlations

The morphological appearance of the surface epithelium of UTJ strongly suggests an active physiological role in the reproductive processes. The secretory activity in the interstitial portion supports the view that tubal fluid is not merely a

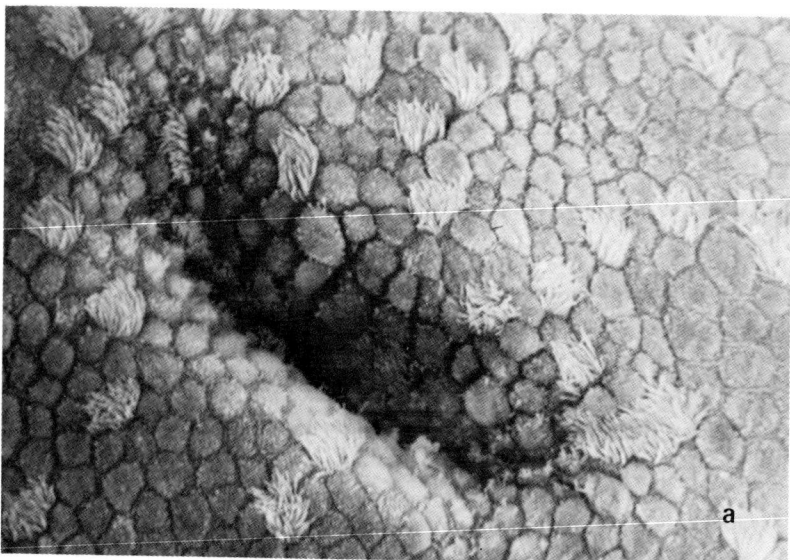

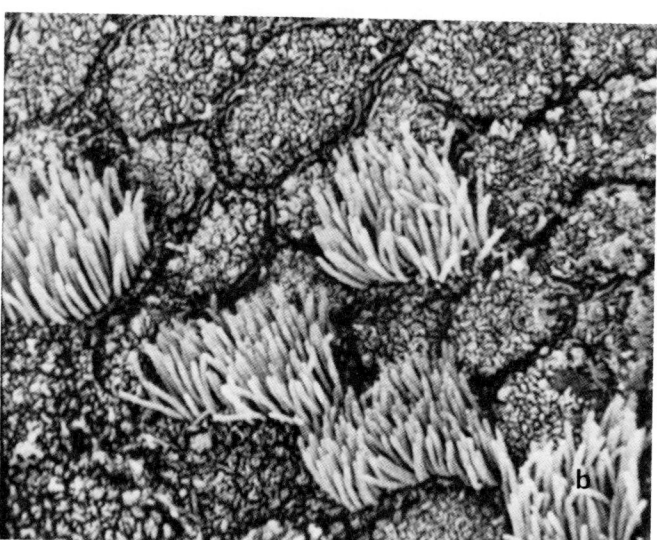

Figure 9.3. (a) Cornual endometrium: postmenopause (X1000). The endometrial cells are flattened and polygonal. The enometrial gland opening is wide and more promininent than in the fertile period (compare with Figure 9.1c).

(b) Cornual endometrium: postmenopause (X3000). The secretory (nonciliated) cells are flattened, but still their surfaces are covered with regular, evenly distributed, rounded, prominent microvilli. The ciliated cells are also well developed with no signs of degeneration or atrophy.

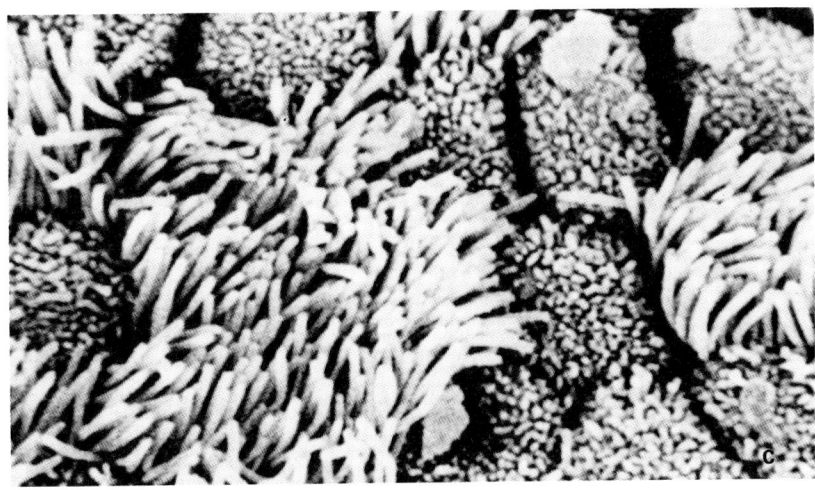

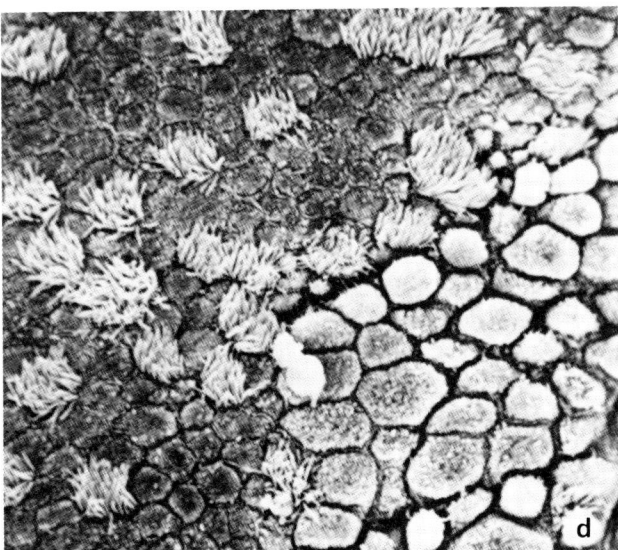

Figure 9.3. (c) Interstitial endosalpinx: postmenopause (X5000). Note the well-preserved ciliated and nonciliated cells with no signs of degeneration or atrophy. The cells are, however, flattened polygonal, the intercellular spaces are markedly widened, and there is no evidence of secretory activity.
(d) The transitional area: postmenopause (X1000). There is a relative increase in the number of ciliated cells. The nonciliated cells are flattened, polygonal and widely separated from each other.

transudate (Brackett and Mastroianni, 1974). These secretions are most abundant at the time when the ovum is still in the tube (early secretory phase). It is reasonable to assume that these secretions contain nutrients that support the ovum and/or spermatozoa. The abundance of ciliated cells in UTJ and specially in the transitional area suggests the presence of strong fluid currents that are probably important in the transport of the spermatozoa into the tube and/or the ova into the uterine cavity. The muscular wall of the intramural tube may under neurohormonal influences actually direct the developing morula to its site of implantation. This muscular mechanism may also act to prevent retrograde menstruation or reflux of lochia or other debris from the postpartum uterus into the tube. Hafez (1973) has presented a detailed discussion of the function of the mammalian UTJ.

CONCLUDING REMARKS

The structure of the surface epithelium of the interstitial portion of the fallopian tube and its merging with the cornual endometrium as seen by SEM has been described. The similarity of the component cells, both secretory and ciliated, in the two epithelia has been noted. A marked increase in the number of ciliated cells and the flattening of the secretory cells at the transitional area has been observed. Structural changes during the different phases of the menstrual cycle and in the postmenopausal period have been described. The secretory and ciliary activities noted support an active role of the UTJ in various reproductive processes. This makes this area a potential target for achieving female sterilization through a transcervical approach, either by occlusion of the lumen or by destroying the surface epithelium by cautery, cryosurgery, or instillation of caustics.

ACKNOWLEDGMENTS

The author acknowledges the collaboration of his co-workers in this study: Dennis Berns, B.S., George D. Wilbanks, M.D., F.A.C.G., Erich E. Brueschke, M.D., and Lourens J. D. Zaneveld, D.V.M., Ph.D., from the Department of Obstetrics and Gynecology, Rush Medical College, Rush Presbyterian-St. Luke's Medical Center, Chicago, Illinois 60612, and Medical Sciences and Engineering Division, Illinois Institute of Technology Research Institute, Chicago, Illinois 60616.

REFERENCES

Brackett, B. G., and L. Mastroianni, Jr. "Compositon of Oviduct Fluid," in *The Oviduct and Its Functions,* A. D. Johnson and C. W. Foley, Eds. (New York: Academic Press, 1974), pp. 133-159.

Brenner, R. M., and R. G. W. Anderson. "Endocrine Control of Ciliogenesis in the Primate Oviduct," in *Handbook of Physiology,* Section 7 (Endocrinology), Vol. II, Part 2, R. O. Greep and E. B. Astwood, Eds. (Washington, D. C.: American Physiological Society, 1973), pp. 123-140.

Brueschke, E. E., *et al.* "Transcervical Tubal Occlusion Using a Steerable Hysteroscope, Implantation of Devices into Extirpated Human Uteri," *Am. J. Obstet. Gynecol.* 127:118 (1977).

David, A., and B. Czernobilsky. "A Comparative Histologic Study of the Uterotubal Junction in the Rabbit, Rhesus Monkey and the Human Female," *Am. J. Obstet. Gynecol.* 101:417 (1968).

Fadel, H. E., *et al.* "Light Microscopic Study of the Epithelium of the Human Uterotubal Junction" (In preparation).

Fadel, H. E., *et al.* "The Surface Epithelium of the Human Uterotubal Junction. A Scanning Electron Microscope Study," in *Scanning Electron Microscopy, 1976,* O. Johari and Becker, Eds. (Chicago: Illinois Institute of Technology Research Institute, 1976a), pp 367-372.

Fadel, H. E., *et al.* "The Human Uterotubal Junction: A Scanning Electron Microscope Study During Different Phases of the Menstrual Cycle," *Fertil. Steril.* 27:1176 (1976b).

Ferenczy, A., and R. M. Richart. "Scanning and Transmission Electron Microscopy of the Human Endometrial Surface Epithelium," *J. Clin. Endocrinol. Metab.* 36:999 (1973).

Ferenczy, A., and R. M. Richart. "Scanning Electron Microscopy of Human Female Genital Tract," *N. Y. State J. Med.* 74:794 (1974).

Gaddum-Rosse, P., R. E. Rumery, R. J. Blandau, and J. B. Thiersch. "Studies on the Mucosa of Post-Menopausal Oviducts: Surface Appearance, Ciliary Activity, and the Effect Estrogen Treatment," *Fertil. Steril.* 26:951 (1975).

Hafez, E. S. E. "Anatomy and Physiology of the Mammalian Uterotubal Junction," in *Handbook of Physiology,* Section 7 (Endocrinology) Vol. II, Part 2, R. O. Greep and E. B. Astwood, Eds. (Washington, E. D.: American Physiological Society, 1973), pp 87-96.

Hafez, E. S. E., and D. L. Black. "The Mammalian Uterotubal Junction," in *The Mammalian Oviduct,* E. S. E. Hafez and R. J. Blandau, Eds. (Chicago: University of Chicago Press, 1969), pp. 85-126.

Johannisson, E., and L. Nilsson. "Scanning Electron Microscopic Study of the Human Endometrium," *Fertil. Steril.* 23:613 (1972).

Lisa, J. R., J. D. Gioia, and I. C. Rubin. "Observations on the Interstitial Portion of the Fallopian Tube," *Surg. Gynecol. Obstet.* 99:159 (1954).

Patek, E. "The Epithelium of the Human Fallopine Tube. A Surface Ultrastructural and Cytochemical Study," *Acta Obstet. Gynecol. Scand.* 53: Suppl 31 (1974).

Patek, E., L. Nilsson, and E. Johannisson. "Scanning Electron Microscopic Study of the Human Fallopian Tube. Report II. Fetal Life, Reproductive Life, and Postmenopause," *Fertil. Steril.* 23:719 (1972).

Rocker, I. "The Anatomy of the Utero-Tubal Junction Area," *Proc. Soc. Med.* 57:707 (1964).

Rumery, R. E., and E. M. Eddy. "Scanning Electron Microscopy of the Fimbriae and Ampullae of Rabbit Oviducts," *Anat. Rec.* 178:83 (1974).

Schueller, E. F. "Ultrastructure of Ciliated Cells in the Human Endometrium," *Obstet. Gynecol.* 41:188 (1973).

Sweeney, W. J., III. "The Interstitial Portion of the Uterine Tube—Its Gross Anatomy, Course and Length," *Obstet. Gynecol.* 19:3 (1962).

CHAPTER 10

THE CERVIX UTERI

Josephine M. Allen and J. A. Jordan

The cervix is a cylindrical structure which forms the lower part of the uterus and provides a tubular connection between the uterus and the vagina (Figure 10.1). It is covered internally and externally with three types of epithelium: original columnar, original squamous and metaplastic squamous. The columnar epithelium lining the endocervical canal is continuous with the epithelium of the endometrium, and the squamous epithelium of the ectocervix is continuous with that of the vagina. The junction between the columnar and squamous epithelia is normally well defined, and may be positioned at any point across the endo- or ectocervix. Under certain physiological conditions, however, the junction becomes a site of a process called squamous metaplasia, whereby columnar epithelium that has been everted onto the ectocervix is replaced by squamous epithelium. This process has been described in detail by Coppleson and Reid (1967), who postulated that cervical neoplasia may be initiated during the process of metaplasia.

The ultrastructure of cervical epithelium has been extensively studied by SEM (Ferenczy and Richart, 1973; Jordan, 1976; Jordan and Williams, 1971; Williams et al., 1973), primarily because the cervix is a site for one of the most common forms of cancer in women. In addition, cervical epithelium is easily accessible to examination and biopsy and has, therefore, become a most important means by which squamous carcinoma, especially in its preclinical stage, may be studied (Coppleson and Reid, 1967; Shingleton et al., 1968; Shingleton and Lawrence, 1976).

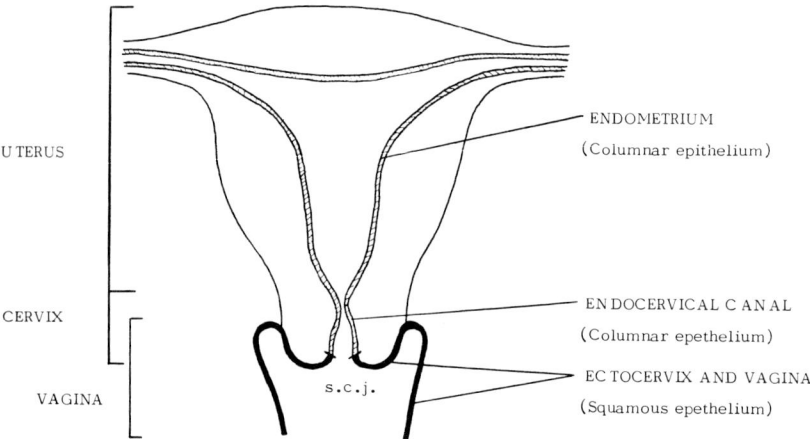

Figure 10.1. Diagrammatic representation of the relationship between the uterus, cervix and vagina: scj, squamo-columnar junction.

THE SURFACE ULTRASTRUCTURE OF CERVICAL EPITHELIUM

Original Columnar Epithelium

Columnar epithelium from the endocervical canal has a dissected appearance due to the presence of finger-like villi and ridges separated by deep clefts, which are oriented obliquely toward the external os. These structures are covered with elongated columnar cells (Figure 10.2a). Only the tips of these small (about 4μm in diameter) closely packed cells are visible and their irregular arrangement has a cobblestone-like appearance (Figure 10.2b). Two types of cells are present: secretory columnar cells and ciliated columnar cells. The surface structure of the secretory cells consists of numerous microvilli, which are about 2 μm in length (Figure 10.3a). No intercellular junctions are seen between the cells. Interspersed with the secretory columnar cells are ciliated columnar cells (Figure 10.3b). The function of these cells is uncertain, although it is believed that they produce movement of mucus.

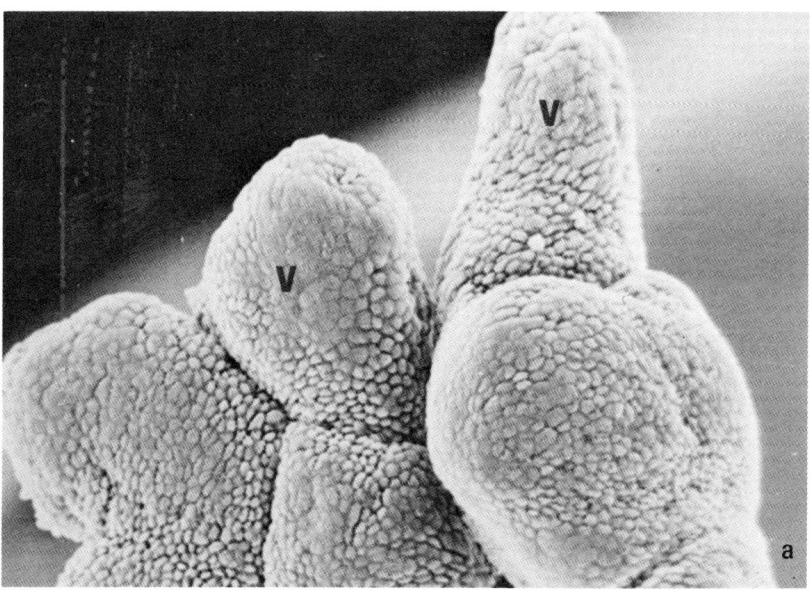

Figure 10.2. (a) Columnar epithelium from the human cervix with numerous finger-like villi: v, villi (X268). (b) Columnar cells from the human cervix. Only the tips of these elongated cells are visible (X2400).

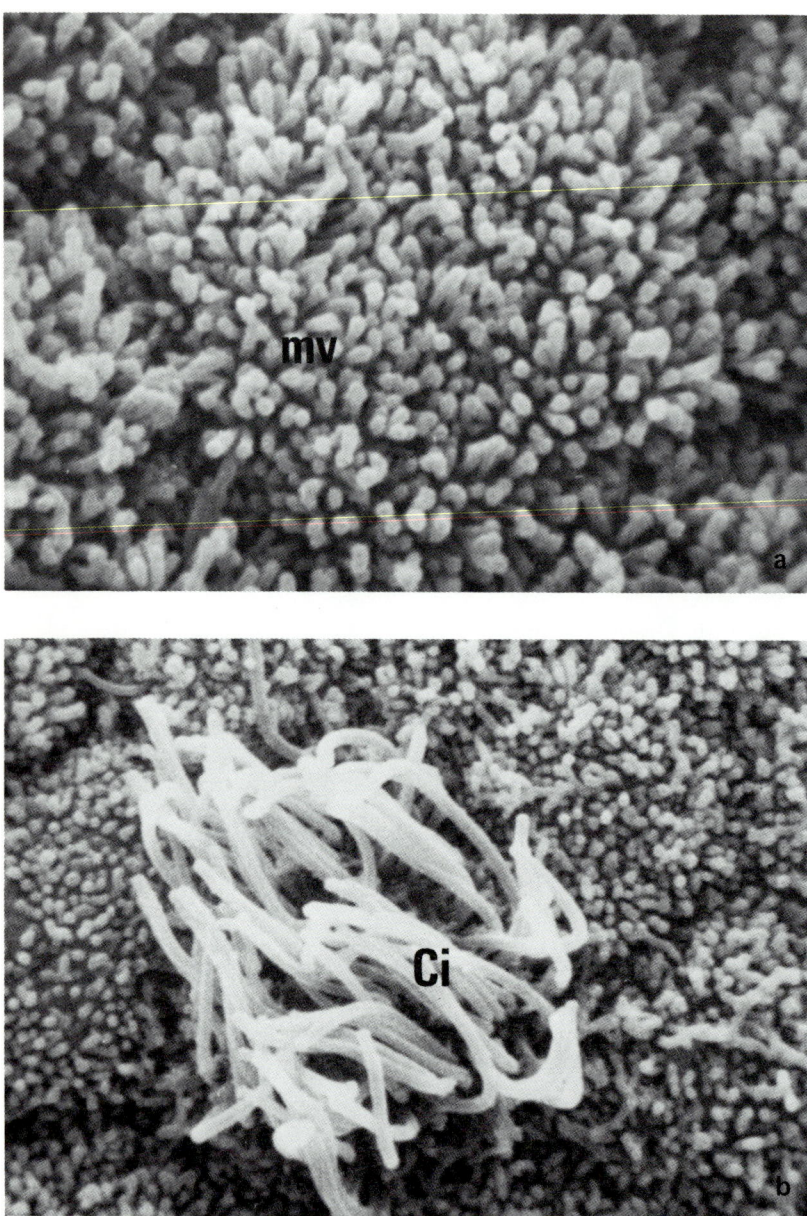

Figure 10.3. (a) The tip of a secretory columnar cell from the human cervix. The surface consists of numerous microvilli: mv, microvilli (X5300). (b) A ciliated columnar cell surrounded by secretory columnar cells from the human cervix: Ci, cilia (X4000).

Original Squamous Epithelium

When examined at low magnification, squamous epithelium has a smooth appearance, by contrast to the dissected appearance of columnar epithelium (Figure 10.4a). Only the uppermost layer of cells is visible and these cells are large (30-40 μm) in diameter, flat and polygonal with raised boundaries between adjacent cells (Figure 10.4b). At higher magnification, the surfaces of the individual cells are seen to consist of a microridge structure (Figure 10.5a). These microridges vary in length, but are approximately 0.15 μm wide with spaces between them of about 0.25 μm. They do not appear to be oriented in a particular way, but are sometimes seen arranged parallel to the cell boundary. The function of these structures is uncertain, but they may help to maintain the integrity of the epithelium by interdigitating, thus offering resistance to sideways movement between layers of cells.

Metaplastic Squamous Epithelium

Metaplasia is a process by which columnar epithelium is converted to squamous epithelium. Three stages of the metaplastic process have been described colposcopically by Coppleson and Reid (1967), and these stages may also be identified with the SEM (Williams *et al.*, 1973). Coppleson and Reid described the first stage of metaplasia as a shortening of the columnar cells at the tips of the columnar villi and the appearance of a cuboidal-type cell. With the SEM this stage is identified by the appearance of larger cells among the regular columnar cells at the tips of the columnar villi (Figure 10.5b). The surfaces of these cells are slightly rounded and covered with short, closely packed microvilli. No terminal bars are visible between the cells, but the cell surface consists of a microvillous structure (Figure 10.6a). In the final stages, when metaplasia is complete, the new squamous epithelium may be seen to consist of mature cells with a microridge surface structure.

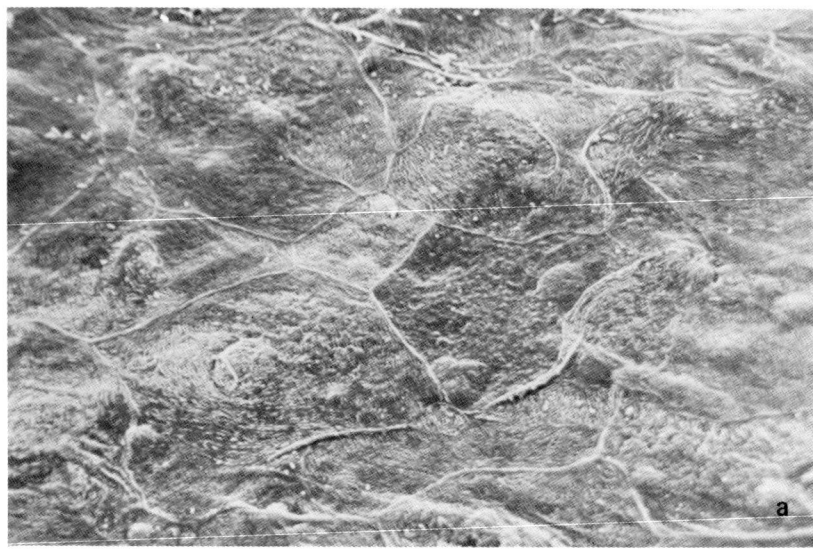
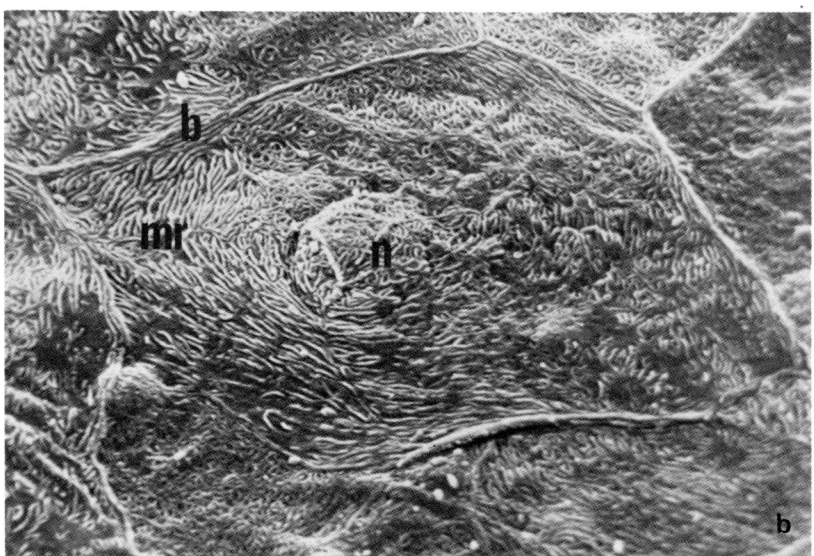

Figure 10.4. (a) Mature squamous epithelium of the human cervix, consisting of large flat polygonal cells which overlap one another (X700). (b) Detail from Figure 10.4a showing one squamous cell. The surface of the cell consists of microridges, some of which are parallel to the cell boundary. The nucleus is also visible: mr, microridges; b, cell boundary; n, nucleus (X1680). (Reproduced by permission from *Cancer Res.*)

Gynecology 153

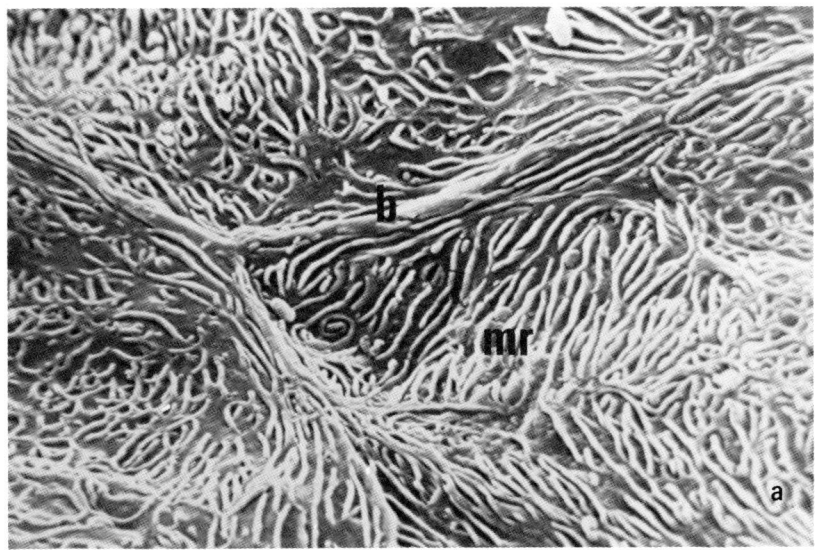

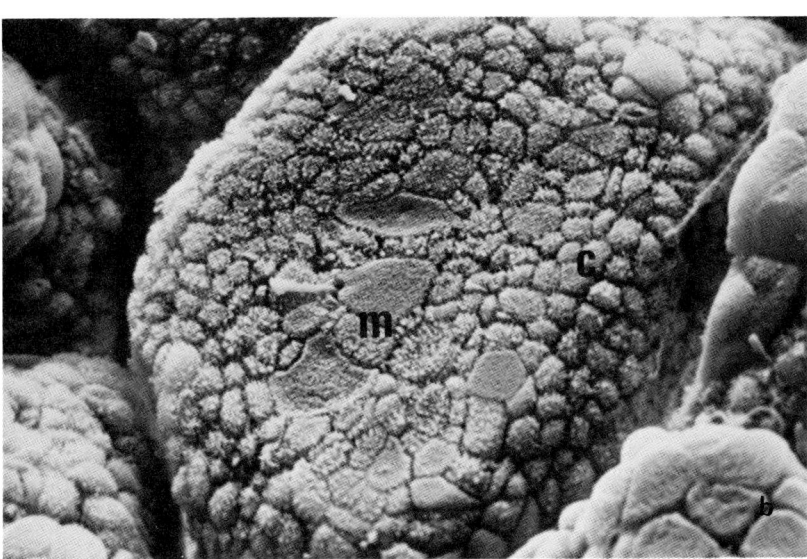

Figure 10.5. (a) Detail from Figure 10.4b showing the junction of three cells with prominent cell boundaries and surface microridges: b, cell boundary; mr, microridges (X3900). (Reproduced by permission from *Cancer Res.*) (b) Columnar villous from the human cervix showing the early stages of metaplasia, where larger more flattened cells are visible among the columnar cells: m, metaplastic cells; c, columnar cells (X650).

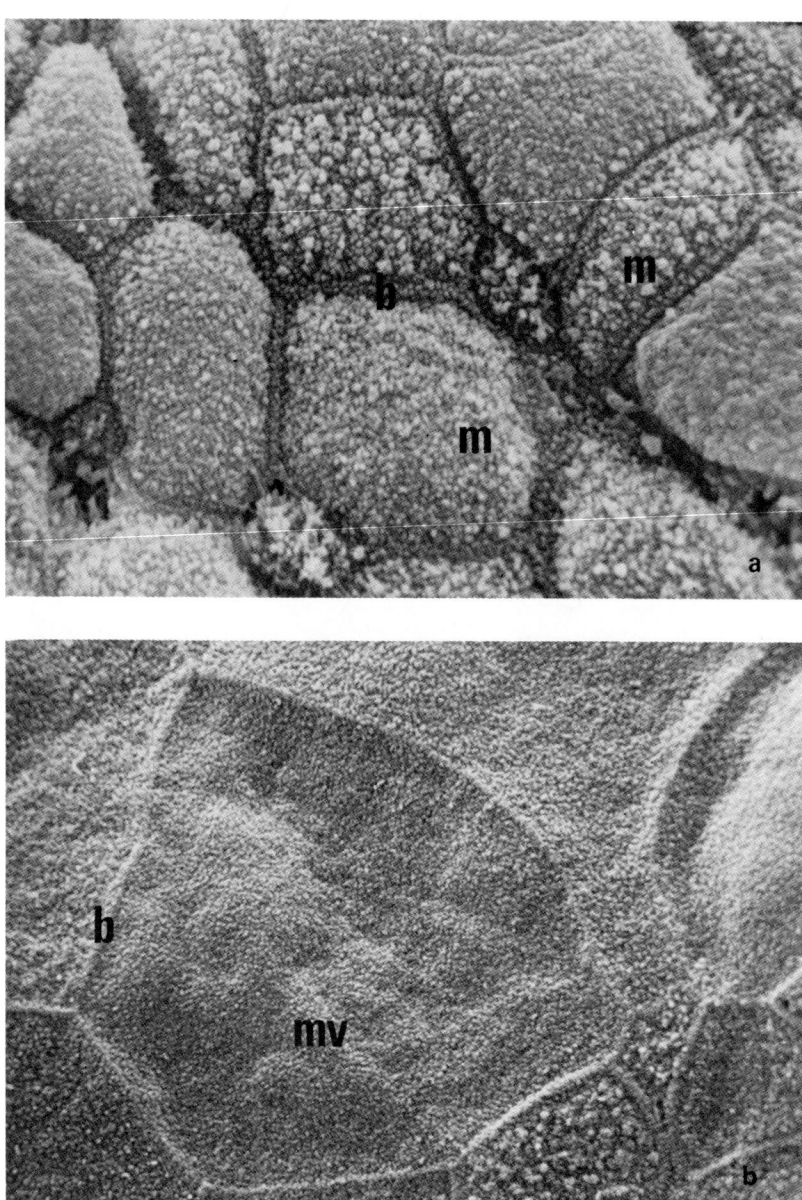

Figure 10.6. (a) Island of metaplastic cells formed by the fusion of columnar villi: m, metaplastic cells; b, cell boundary (X1650). (b) Large flattened polygonal cell seen in the later stages of metaplasia. The cell has prominent cell boundaries and the surface consists of numerous short microvilli: b, cell boundary; mv, microvilli (X2200).

Gynecology 155

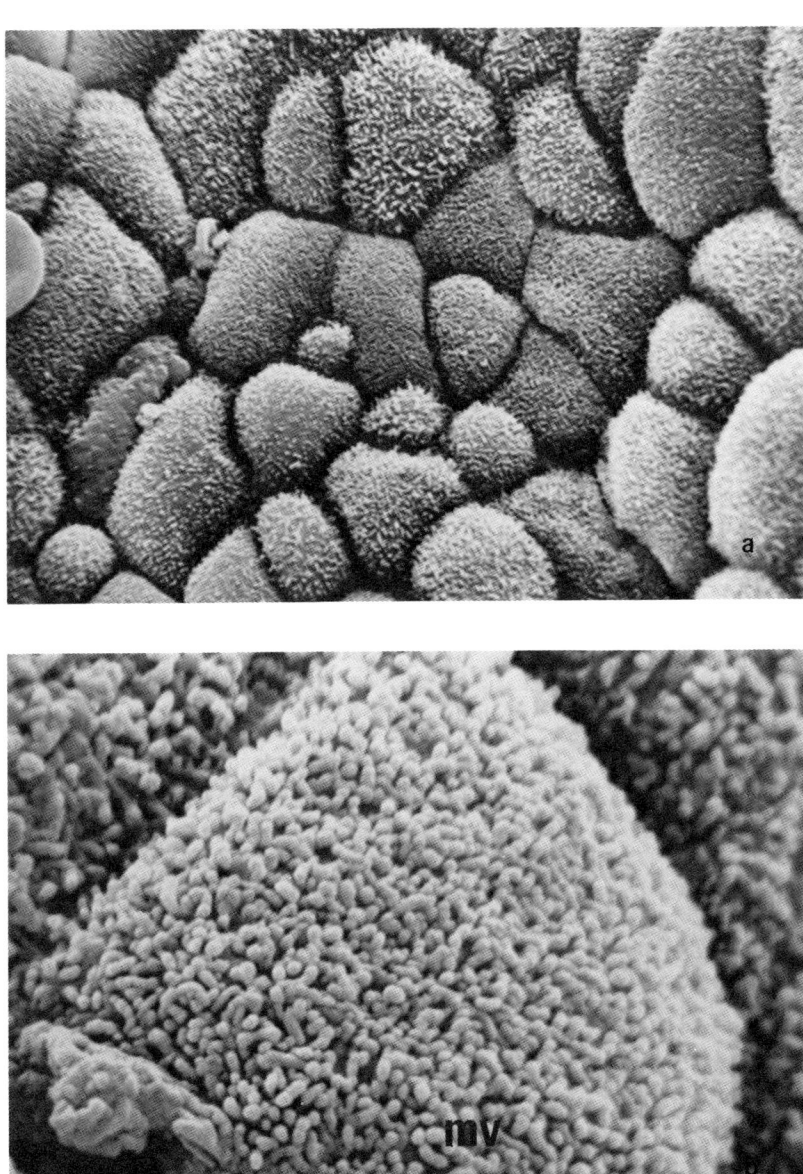

Figure 10.7. (a) Carcinoma-*in-situ* in human cervical epithelium, with a disorganized appearance due to the irregular shape and size of the cells (X2500). (b) Detail of carcinoma-*in-situ* cell, which is small and rounded with a surface consisting of numerous microvilli: mv, microvilli (X7700).

Carcinoma-*In-Situ*

Carcinoma-*in-situ* (CIS) differs quite markedly from the appearance of normal epithelium. When examined at low magnification, it has a disorganized appearance with an uneven surface by contrast to the appearance of normal squamous epithelium (Figure 10.7a). The individual cells are rounded and of irregular shape and size, with no well-defined intercellular junction. The surfaces of the individual cells are covered with numerous microvilli, which are about 0.15 μm in diameter (Figure 10.7b). The presence of these microvilli in addition to the rounded shape of the cells may inhibit interdigitation of adjacent surfaces, thus leading to the increased exfoliation of these cells.

CONCLUDING REMARKS

Examination of cervical epithelium by SEM has demonstrated morphological and ultrastructural features not previously shown by other techniques. The appearance of the cytoplasmic membrane of mature squamous cells has been described by Transmission Electron Microscopy as being microvillous (Shingleton and Lawrence, 1976), although with the SEM a microridge structure is observed. It is possible that these structures interdigitate more efficiently than microvilli, thus helping to maintain the integrity of the epithelium and its resistance to mechanical trauma. By contrast, metaplastic and CIS cells have a microvillous surface structure, which is probably a reflection of the degree of activity within the cell. These structures have also been described by Transmission Electron Microscopy (Shingleton and Lawrence, 1976). The correlation of the SEM observations with those of colposcopy and histology of the same area of epithelium is possible and has also formed an important part of the study of the human cervix with SEM (Ayres *et al.*, 1971; Murphy *et al.*, 1974).

REFERENCES

Ayres, A., J. M. Allen and A. E. Williams. "A Method of Obtaining Conventional Histological Sections from Specimens after Examination by Scanning Electron Microscopy," *J. Micros.* 93:247-250 (1971).

Coppleson, M. and B. Reid. *Pre-clinical Carcinoma of the Cervix Uteri* (London: Pergamon Press, 1967).

Ferenczy, A. and R. M. Richart. "Scanning Electron Microscopy of the Cervical Transformation Zone," *Am. J. Obstet. Gynecol.* 115:151-157 (1973).

Jordan, J. A. In: *The Cervix*, J. A. Jordan and A. Singer, Eds. (London: W. B. Saunders, 1976), pp. 44, 372.

Jordan, J. A. and A. E. Williams. "Scanning Electron Microscopy in the Study of Cervical Neoplasia," *J. Obstet. Gynaecol. Brit. Commonwlth.* 78:940-946 (1971).

Murphy, J. F., J. A. Jordan, J. M. Allen and A. E. Williams. "Correlation of Scanning Electron Microscopy, Colposcopy and Histology in 50 Patients Presenting with Abnormal Cervical Cytology," *J. Obstet. Gynaecol. Brit. Commonwlth.* 81:236-241 (1974).

Shingleton, H. M., R. M. Richart, J. Weiner and D. Spires. "Human Cervical Intra-Epithelial Neoplasia: Fine Structure of Dysplasia and Carcinoma-*In-Situ*," *Cancer Res.* 28:695-706 (1968).

Shingleton, H. M. and D. Lawrence. In: *The Cervix*, J. A. Jordan and A. Singer, Eds. (London: W. B. Saunders, 1976), pp. 36, 363.

Williams, A. E., J. A. Jordan, J. M. Allen and J. F. Murphy. "The Surface Ultrastructure of Normal and Metaplastic Cervical Epithelium and of Carcinoma-*In-Situ*," *Cancer Res.* 33:504-513 (1973).

CHAPTER 11

CERVICAL MUCUS

L. J. D. Zaneveld, P. F. Tauber, D. Propping and E. S. E. Hafez

The rheological parameters of cervical mucus are determined by the biophysical arrangement of the network, the diameter of backbone fibers, dimensions of secondary and tertiary microfibrils, and arrangement of aqueous intermicellar cavities. These, in turn, are influenced by the stage of the menstrual cycle, hormonal milieu, distance from the cervical crypts, and biochemical and biophysical interactions with the endometrial and vaginal fluids. Under the influence of ovarian hormones, cervical mucus undergoes dramatic quantitative, biophysical and biochemical changes throughout the menstrual cycles. During midcycle, coinciding with the time of ovulation, cervical mucus is most favorable for survival and transport of spermatozoa. The duration of this "favorable phase" for sperm penetration varies among individual women.

RHEOLOGICAL CHARACTERISTICS

Once secreted by the cervical crypts, which are at right angles to the cervical canal, the mucus is released perpendicular to the axis of that channel (Figure 11.1). Most likely due to the force of gravity, the filaments flow down parallel to the axis. The cervical canal thus acts as a capillary tube where the main filaments have a parallel arrangement. Once the column of

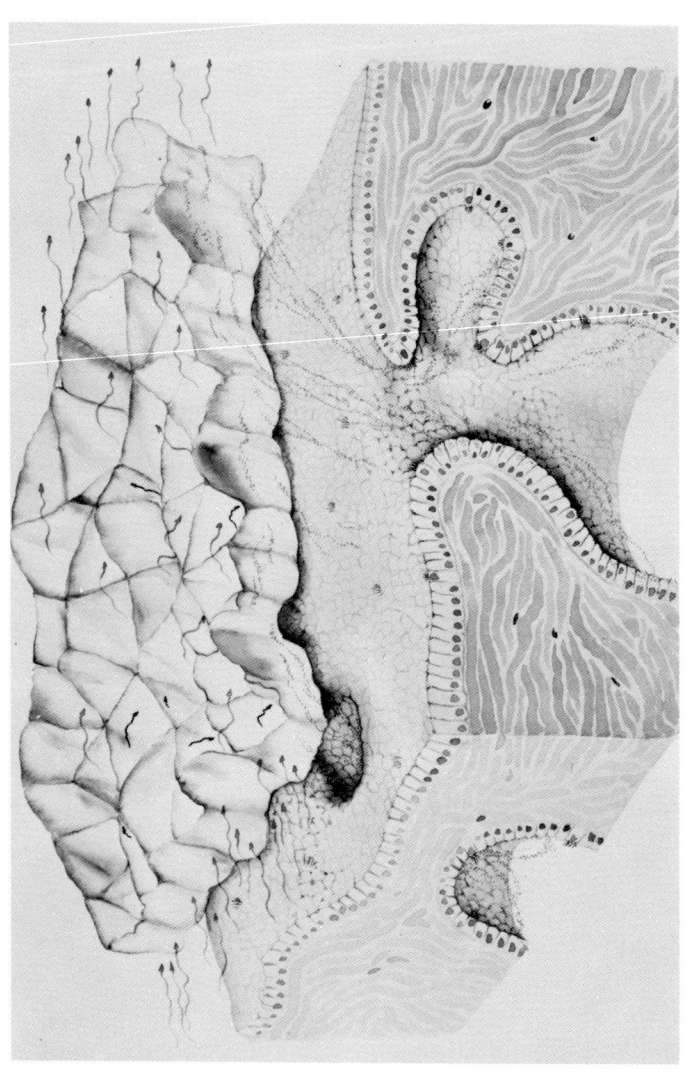

Figure 11.1. Diagram of the production of cervical mucus (honeycomb structure shown at the top). Macromolecules of glycoprotein (dots) are elaborated by the cyclitory cells in the cervical crypts to form the mesh of cervical mucus. A few ciliated cells are shown within the crypts and at the openings of the crypts. During midcycle the mesh of the cervical mucus is quite wide to allow the transport of sperm from the site of ejaculation to the uterine lumen. It is possible that the micelles of the cervical mucus filters the ejaculate when nonviable spermatozoa (dark color) adhere to the mucus.

cervical mucus flows out of the external os, it is suspended within the vaginal fluid accumulating in the vaginal fornix. There the cervical mucus is freed of the capillary biophysical constraints encountered in the canal. After intercourse, and entry into the cervical mucus, the spermatozoa are protected from the acidic pH of the vaginal milieu and are transported rapidly between the micelles of mucus within the vaginal pool.

The cervical mucus gel consists of a network of thread-like flexible macromolecules of glycoprotein known as mucins (Elstein et al., 1973). The rheological properties of cervical mucus depend upon the orientation of these filamentous macromolecules. The glycoprotein strands are linked together by oblique or transversal bonds and organized into microfibrils and fibers. Often they form a honeycomb macromolecular arrangement. The basic organic material tends to be arranged in a coiled fashion (Chrétien et al., 1975). Utilizing biophysical techniques, Odeblad (1968) postulated that in midcycle, under estrogenic influence, the cervical mucin molecules group together in micelles (bundles) of 0.5 μm in diameter. Aqueous cavities of up to 10 μm between these bundles contain the cervical plasma in which there is movement of fluids and through which spermatozoa are transported. Under a progestational dominance there is little or no micelle formation and the macromolecular filaments do not have directional or parallel arrangement, forming a tight meshwork of small mesh size (0.3 μm). This has been confirmed by scanning electron microscopy.

Fresh cervical mucus is rheologically heterogeneous reflecting compositional differences *in situ*. The viscoelasticity of cervical mucus, which depends on the concentration of nondialyzable solids, is relatively stable when samples are stored at ambient temperature (Wolf et al., 1977a).

When the contribution of nondialyzable solids to viscoelasticity is minimized by data normalization or by sample reconstitution, a significant increase in viscoelasticity is associated with the ovulatory phase suggesting the occurrence of a relative increase in mucin concentration or a compositional change in the mucus (Wolf et al., 1977b). Cyclic fluctuations in the viscoelastic properties of cervical mucus during midcycle reflect alterations in the concentration of macromolecules in the mucus in response to the changing hormonal milieu (Wolf *et al.*, 1977c).

SCANNING ELECTRON MICROSCOPY OF MUCUS

Transmission electron microscopy is not particularly suitable for observing spatial distribution of the gel macromolecules of cervical mucus (Singer and Reid, 1970). Extensive investigations using SEM have been conducted on the ultrastructure of cervical mucus throughout the menstrual cycle.

The method of preparation of the cervical mucus sample is very critical since the spatial arrangement of the filaments differs considerably whether or not it has been stretched. For example, because of its extreme thinness, the marginal area of a droplet of cervical mucus on a glass slide has different biophysical properties than the center. The middle portion of the sample is thick and less exposed to the biophysical constraints. After dehydration, cervical mucus appears composed of a framework of interlacing filaments, thus forming a network whose meshes vary in size at different phases of the menstrual cycle.

The macromolecules make up microfibrils 200-1500 Å in diameter that are arranged into larger fibers of 0.5 and 15 μm in diameter. Homogenized samples of normal cervical mucus consist of globular particles 1000-1500 Å in diameter in the proliferative phase. The globules may be linked together by thinner strands (Elstein *et al.*, 1971). In early luteal phase, globular structures are found together with fibrils 200 Å in diameter and 3000-5000 Å in length.

The cervical mucus consists of a three-dimensional membranous honeycomb-like structure (Figures 11.2 and 11.3). The cavities within this honeycomb structure increase from 2-6 μm in the early proliferative phase to 30-35 μm in the late proliferative phase, at which time spermatozoa can pass through the channels (Daunter *et al.*, 1976). In the luteal phase there is an increase in cellular debris, and the membranous honeycomb network becomes compact with pore sizes similar to those in the early proliferative phase (Elstein and Daunter, 1976).

Cervical Mucus at Midcycle

Throughout the midcycle period cervical mucus contains microfibrils measuring 800 Å in width with a range between 500-1500 Å (Zaneveld et al., 1975a).

On the day of the LH surge, a multitude of microfibrils interconnect making larger macrofibrils (micelles) with variable diameter (Figure 11.4). Large fibers measure 2-5 μm in diameter whereas small fibers measure 0.5-1 μm in diameter (Zaneveld et al., 1975a).

The parallel arrangement and the length of the fibers give the spermatozoa their directional approach toward the uterus or the cervical crypts during their transport through the cervix. During the ovulatory phase, the mesh attains maximal size and the dimensions of the channels may exceed 10 μ. The diameter of channels among micelles of glycoproteins reaches a maximum on the day of ovulation. During the luteal phase, there is a denser, finer mesh of these macromolecules with small channels between the fibrils which form a barrier between the vagina and the upper genital tract.

Crystallization of Cervical Mucus

Remarkable changes occur in the ferning pattern of air-dried cervical mucus throughout the menstrual cycle. Ferning (arborization) reaches a peak at midcycle or after estrogen treatment, whereas progestational agents prevent fern formation. Different ferning patterns are present during pregnancy and menopause (Roland, 1958). Sperm transport through the cervical mucus is under hormonal control, and a direct relation exists between the optimal time of sperm penetration and the degree of ferning (Moghissi and Marks, 1971). Several attempts have been made to "grade" the fern test for clinical evaluation of hormonal state and to determine the optimal time for artificial insemination (Kesseru, 1972).

Different types of crystal configuration and aggregation are noted during midcycle using scanning electron microscopy (Table 11.1) (Zaneveld et al., 1975b). The time span between the days that long rows of crystals appear (at least in the one individual

164 *Scanning Electron Microscopy of Human Reproduction*

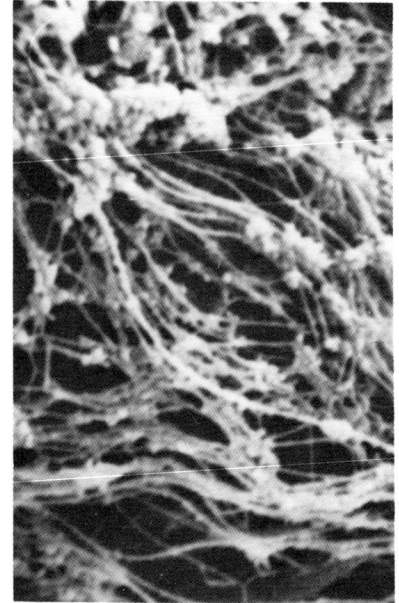

A

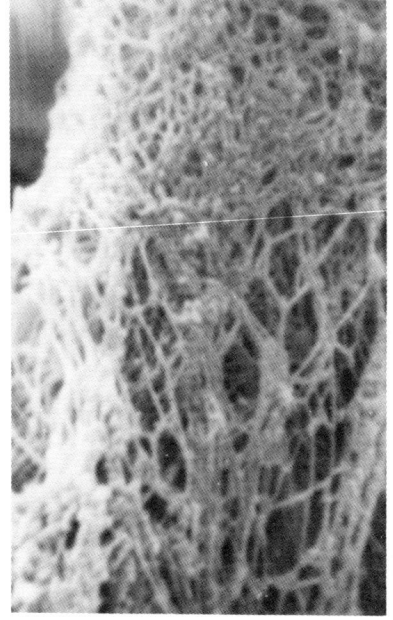

B

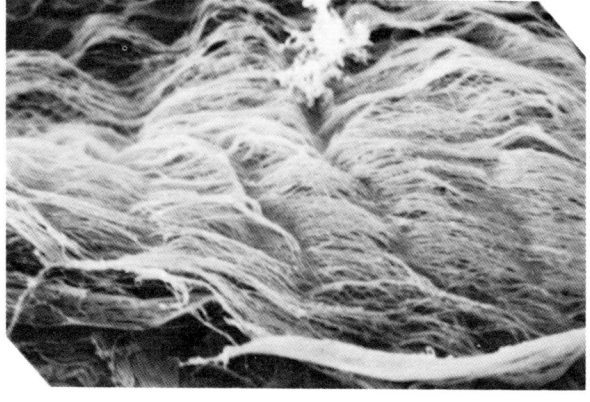

C

Gynecology 165

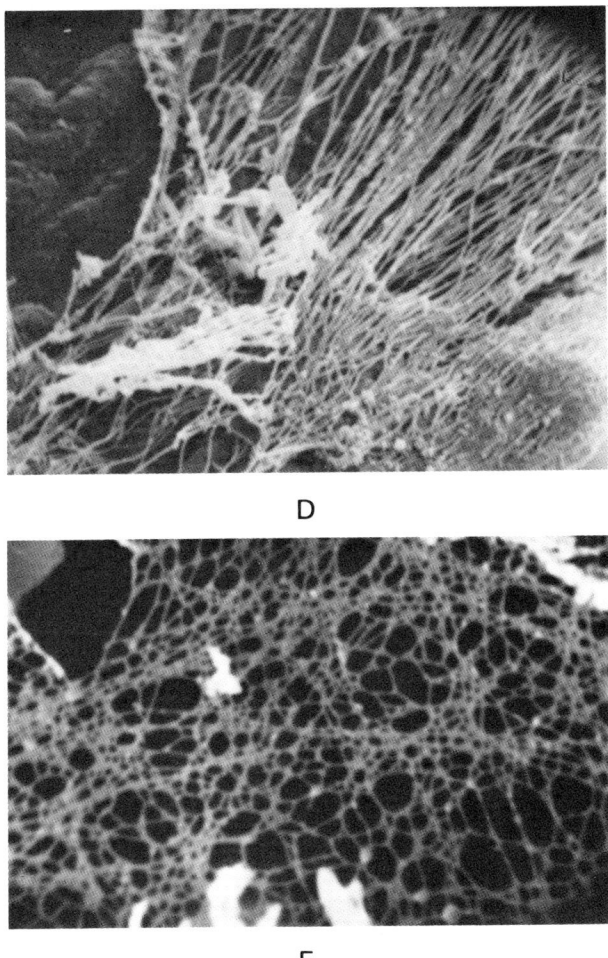

Figure 11.2. Microfibrillar structures of midcycle human cervical mucus. The arrangements of the microfibrils vary from parallel to networks. The microfibrils are always interconnected independent of their arrangement. Thus, if the parallel-arranged microfibrils were stretched perpendicular to their flow, they would form networks: (A) X12,000; (B) X12,000; (C) X3000; (D) X6400; (E) X12,000 (Zaneveld *et al.*, 1975a).

A

Gynecology 167

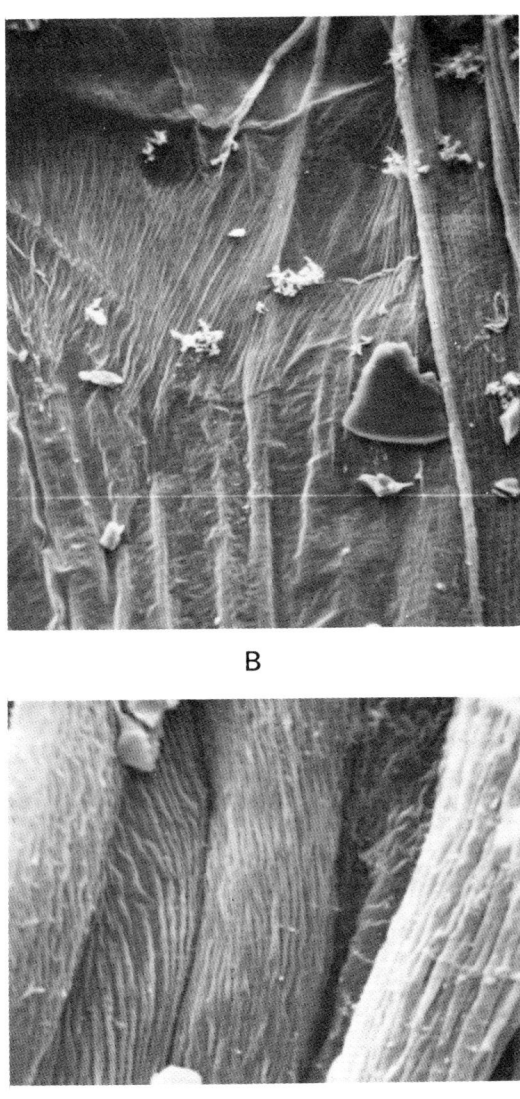

Figure 11.3. Fibers of midcycle human cervical mucus. Microfibrils join to form small fibers which in turn coalesce into large fibers. The surface of the large fibers (Figure 11.2C) shows their fibrillar composition: (A) X2300; (B) X1300; (C) X5800 (Zaneveld et al., 1975a).

Figure 11.4. Crystallization pattern of cervical mucus taken on day 12 of the menstrual cycle: (a) type B fern (X112); (b) type C fern

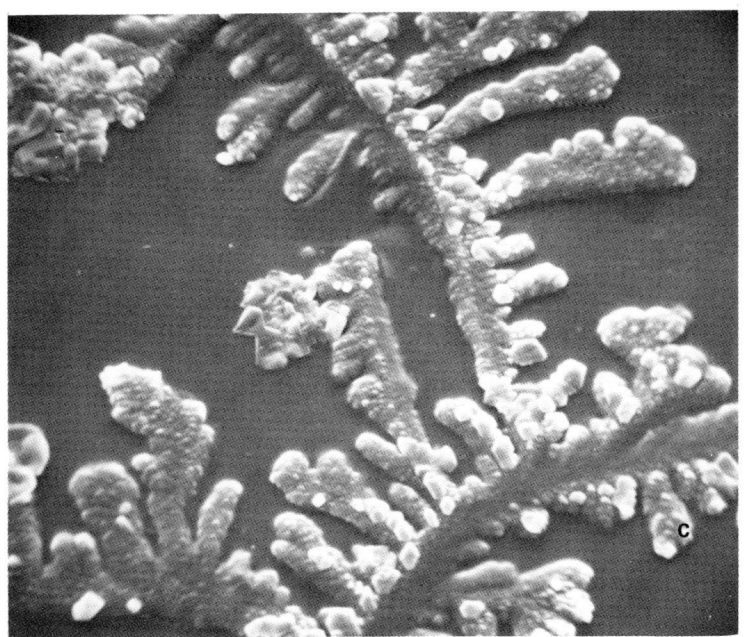

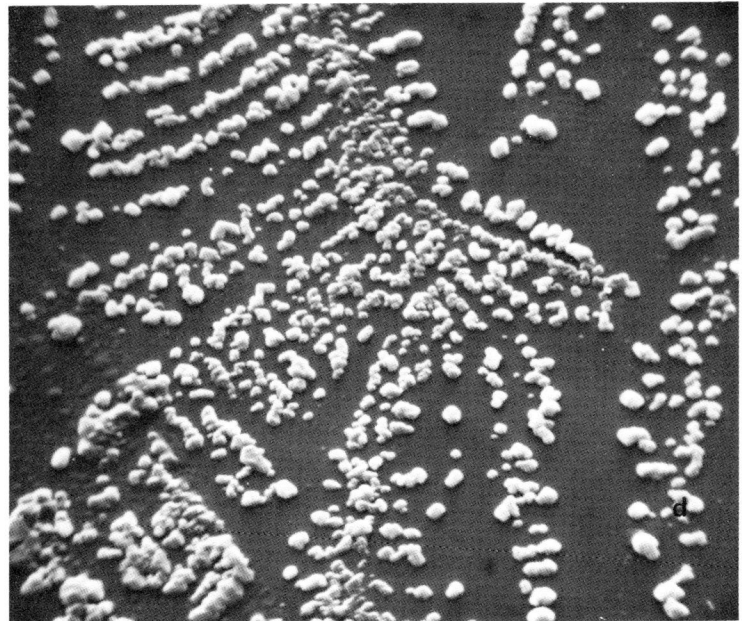

(X250); (c) type B fern (X2400); (d) type A fern (X2500) (Zaneveld et al., 1975a).

Table 11.1

Crystallization Pattern of Human Cervical Mucus as Observed by Scanning Electron Microscopy during Midcycle (Zaneveld et al., 1975b)

Phase of Cycle	Crystallization Pattern on the Strands
Very early follicular phase	No crystals or small groups of crystals
−3 days before ovulation	Long rows of crystals
−2 days before ovulation	Primitive crystalline fern patterns
Presumable ovulation as determined by the peak in LH	More advanced pattern of crystalline fern (type A) change to completely "true" fern form (type C) at the time of ovulation. Intermediate forms B are also formed when drops of mucus are dried
+2 days after ovulation	Type B fern appears as an intermediate form between A and C
+3 days after ovulation	Long rows of crystals
+5 and late luteal phase	Individual crystals sparsely distributed over strands

tested) is exactly six days, ovulation occurring somewhere in the middle. Further studies should be performed to verify this in other patients.

During early stages of the cycle the sodium and chloride ions appear as flat, square crystals whereas calcium and phosphate ions appear as tall, rectangular crystals. Later in the cycle the flat, square crystals assume cuboidal shapes, which are quite typical for NaCl (Figure 11.5). Crystallization patterns during the luteal phase are similar to those of the follicular phase. Changes in the crystallization patterns occur daily, particularly during the midcycle period. Such changes are more apparent on "strands" than on "droplets" of cervical mucus. For clinical evaluation of the crystallization pattern, the use of strands is highly recommended (Davajan et al., 1971). These strands are obtained by drawing a cervical mucus drop out as far as possible with a Pasteur pipette and placing the strand on a slide. Ferning results from the interaction of electrolytes with high-molecular-weight mucin in the cervical mucus (Beck et al., 1971).

FUNCTIONAL SIGNIFICANCE

The cervical mucus plays an essential role in the reproductive process. It acts as a temporary mechanical barrier during most of the ovarian cycle. However, during the ovulatory phase, cervical mucus is freely penetrable by spermatozoa. Sperm penetrability of mucus first appears between the eighth and ninth days of the cycle and increases gradually to attain a maximum at the time of ovulation (Bergman, 1950). Several days after ovulation, a low level of sperm penetrability in cervical mucus is still demonstrable.

The chemical, biophysical and physiological properties of cervical mucus are quite variable during different phases of the ovarian cycle. Its secretion is stimulated by estrogens, whereas progesterone domination causes not only a quantitative decrease in mucus secretion but correlates also with alterations in its chemical and physical properties. A ten-fold increase in the volume of mucus secreted occurs from the end of menstruation to the time of ovulation, just prior to which the percent

172 *Scanning Electron Microscopy of Human Reproduction*

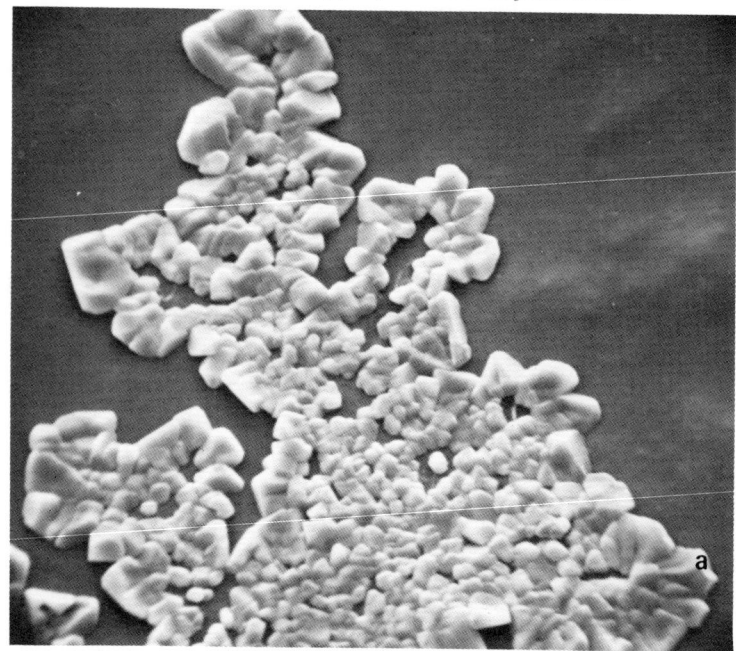

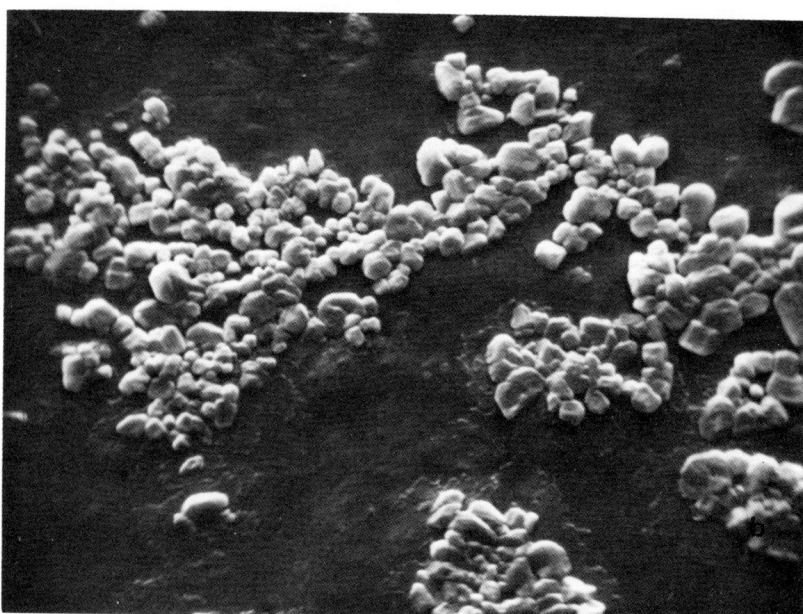

Figure 11.5. Ferning pattern of cervical mucus (X2500): (a) day 9 of the menstrual cycle; (b) day 11 of the menstrual cycle (Zaneveld *et al.*, 1975b).

composition of water increases from a basal level of 92% to approximately 98% water (Pommerenke, 1946). Conversely, from the middle of the cycle to the end of the luteal phase, the percent of solid matter comprising cervical mucus increases from a low of 2% at time of ovulation to a high of 8% prior to menstruation (Odeblad, 1968). The refractive index, viscosity, "Spinnbarkeit," and stickiness of cervical mucus change in direct response to the variations in hormone according to cyclic changes during the menstrual cycle (Cohen *et al.*, 1952; MacDonald, 1969).

Cervical mucus is a hydrogel with a low molecular weight similar to that of diluted serum. Its macromolecular composition resembles that of a mucin. The glycoprotein molecules of mucus are arranged by noncovalent forces into a fibrillar network. One model of the fibrillar structure of cervical mucus proposes that its glycoproteins have chains of oligosaccharides terminated by sialic acid which project laterally from the fibril axis (Moghissi, 1972). However, this model does not account for the variations in the biophysical properties of cervical mucus during the ovarian cycle.

Cervical mucus has a three-dimensional micellar ultrastructure which is consistent with the nuclear magnetic resonance properties of this material (Odeblad, 1968). The meshes of the glycoprotein network enlarge at the time of ovulation to facilitate sperm passage. Conversely, during the luteal phase, the glycoprotein molecules, instead of being organized as predominantly parallel filaments, would be randomly associated as a network with mesh diameters impeding sperm penetration. This model, in which the macromolecules are proposed to be arranged into a network similar to that of a knitted garment or "tricot-like arrangement" can theoretically account for the cyclic variations in the chemical and biophysical properties of cervical mucus. Thus sperm penetrability can be regulated by the opening or the closing of the superstructure of the mucus at the level of fibrillar mesh sizes (Chrétien, 1973). Cervical mucus anomalies account for up to 33-50% of all cases of female sterility (Clavero, 1954; Mazer and Israel, 1963; Simmons and Taymor, 1955; Steinberg, 1958).

REFERENCES

Beck, K. J., U. Budde and A. Neuhaus. "Die kristallographischen Grundlagen des Arboristions-phenomenes des Cervixschleimes," *Arch. Gynakol.* 210:76-96 (1971).

Bergman, P. "Sexual Cycle, Time of Ovulation and Time of Optimal Fertility in Women. Studies on Basal Body Temperature, Endometrium, and Cervical Mucus," in *The Cervical Cycle*, Chapter IV *Acta Obstet. Gynecol. Scand.* 29: Suppl. 4, 74 (1950).

Chrétien, F. C. "L'ultrastructure de la glaire cervicale," *Contracep. Fert. Sexual.* 1:9 (1973).

Chrétien, F. C., J. Cohen and A. Psychoyos. "Human Cervical Mucus during the Menstrual Cycle and Pregnancy in Normal and Pathological Conditions," *J. Reprod. Med.* 14:192 (1975).

Clavero, N. A. "Statistical Study of the Cervix as a Factor in Sterility in 1631 Cases. Results of Treatment," *Rev. Esp. Ostet. Ginec.* 13:266 (1954).

Cohen, M. R., I. F. Stein and B. M. Kaye. " 'Spinnbarkeit:' A Characteristic of Cervical Mucus. Significance at Ovulation Time," *Fertil. Steril.* 3:201 (1952).

Daunter, B., E. Chantler and M. Elstein. "Scanning Electron Microscopy of Cervical Mucus. Normal Menstrual Cycle and Pregnancy," *Brit. J. Obstet. Gynaecol.* 83 (1976).

Davajan, V., R. M. Nakamura and D. R. Mishell. "A Simplified Technique for Evaluation of the Biophysical Properties of Cervical Mucus," *Am. J. Obstet. Gynecol.* 109:1042-1048 (1971).

Elstein, M. and B. Daunter. "The Structure of Cervical Mucus," in *The Cervix*, A. Singer and J. Jordan, Eds., Chapter 11 (London: W. B. Saunders, 1976), pp. 137-147.

Elstein, M., M. Mitchell and J. T. Syrett. "Ultrastructure of Cervical Mucus," *J. Obstet. Gynaecol. Brit. Commonwlth.* 78:180-183 (1971).

Elstein, M., K. S. Moghissi and R. Borth. *Cervical Mucus in Human Reproduction* (Copenhagen: Scriptor, 1973).

Kesseru, E. "A Simple Method for Measuring Crystallization of the Cervical Mucus and its Application in Human Sperm Migration," *Internat. J. Fertil.* 17:201-209 (1972).

MacDonald, R. R. "Cyclic Changes in Cervical Mucus," *J. Obstet. Gynaecol. Brit. Commonwlth.* 76:1090 (1969).

Mazer, C. and S. L. Israel. *Diagnosis and Treatment of Menstrual Disorders and Sterility* (New York: Hoeber, 1963).

Moghissi, K. S. "The Function of the Cervix in Fertility," *Fertil. Steril.* 23:295 (1972).

Moghissi, K. S. and C. Marks. "Effects of Microdose Norgestrel on Endogenous Gonadotropic and Steroid Hormones, Cervical Mucus Properties, Vaginal Cytology and Endometrium," *Fertil. Steril.* 22:424-434 (1971).

Odeblad, E. "The Functional Structure of Human Cervical Mucus," *Acta Obstet. Gynecol. Scand.* 47 (Suppl. I):57-79 (1968).

Pommerenke, W. T. "Cyclic Changes in the Physical and Chemical Properties of Cervical Mucus," *Am. J. Obstet. Gynecol.* 52:1023 (1946).

Roland, M. "The Fern Test: a Critical Analysis," *Obstet. Gynecol.* 11: 30-34 (1958).

Simmons, F. A. and M. L. Taymor. "Failure of Conception in 100 Completely Studied Couples," *Fertil. Steril.* 6:320 (1955).

Singer, A. and B. L. Reid. "The Effect of the Oral Contraceptive Steroids on the Ultrastructure of the Human Cervical Mucus—a Preliminary Communication," *J. Reprod. Fertil.* 23:249-255 (1970).

Steinberg, W. "The Role of Cervical Factor in Fertility," *Fertil. Steril.* 9:436 (1958).

Wolf, D. P., L. Blasco, M. A. Khan and J. Litt. "Human Cervical Mucus. I. Rheologic Characteristics," *Fertil. Steril.* 28:41-46 (1977a).

Wolf, D. P., L. Blasco, M. A. Khan and M. Litt. "Human Cervical Mucus. II. Changes in Viscoelasticity during the Ovulatory Menstrual Cycle," *Fertil. Steril.* 28:47-52 (1977b).

Wolf, D. P., J. Sokoloski, M. A. Khan and M. Litt. "Human Cervical Mucus. III. Isolation and Characterization of Rheologically Active Mucus," *Fertil. Steril.* 28:53-58 (1977c).

Zaneveld, L. J. D., P. F. Tauber, C. Port, D. Propping and G. F. B. Schumacher. "Structural Aspects of Human Cervical Mucus," *Am. J. Obstet. Gynecol.* 122:650-654 (1975a).

Zaneveld, L. J. D., P. F. Tauber, C. Port and D. Propping. "Scanning Electron Microscopy of Cervical Mucus Crystallization," *Obstet. Gynecol.* 46(4):419-428 (1975b).

CHAPTER 12

AMNIOTIC FLUID CELLS*

Sharon M. Noonan and Lester Weiss

Amniotic fluid cells are routinely used today by the obstetrician, neonatologist and pediatrician for *in utero* identification of various physiological or pathological conditions, such as: (1) fetal maturity (Huisjes and Arendzen, 1970); (2) ruptured membranes (Bourgeois, 1942); (3) chromosomal abnormalities (Jacobson and Barter, 1967; Fuchs and Cederquist, 1973); and (4) inborn errors of metabolism (Nadler and Gerbie, 1971). These diagnoses are accomplished by aspiration of amniotic fluid cells from the amniotic sac enclosing the fetus (amniocentesis), followed by the morphologic and/or enzymatic analysis of either cultured or uncultured amniotic fluid cells (Kaback and Leonard, 1972).

The origin of the cells shed into the amniotic fluid are thought to derive from both amnion epithelium (umbilical cord, placental and reflexed amnion) and fetal epithelium (epidermis, digestive and respiratory tract mucosa, and urogenital epithelia). However, questions concerning the definitive identification of the source, quantitative evaluation of the various types of amniotic fluid cells, as well as an understanding of the functional capabilities of specific amniotic cells, remain unresolved.

This chapter concerns our morphological investigation by light microscopy (LM), transmission electron microscopy (TEM)

---

* This investigation was supported in part by a grant from the Detroit General Hospital Research Corporation.

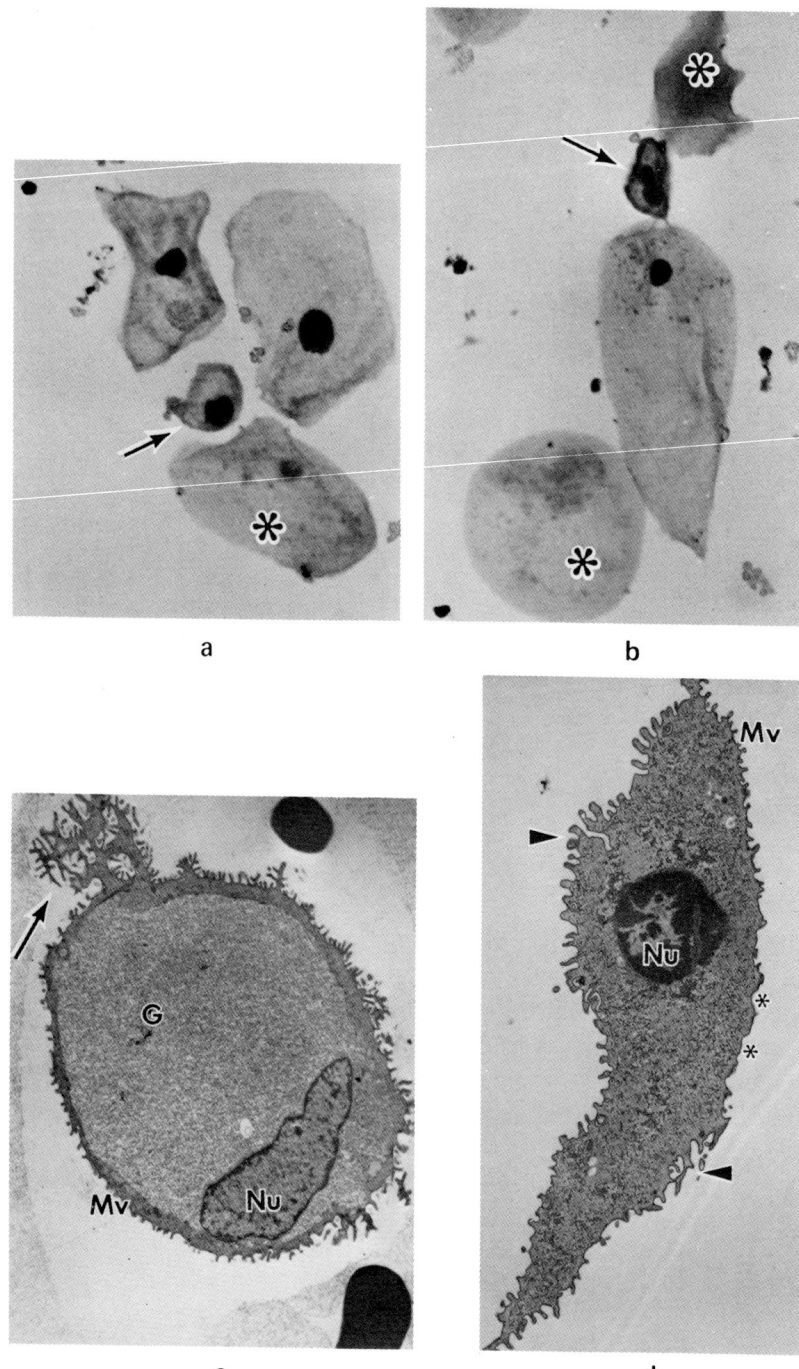

Figure 12.1. Light microscopy and transmission electron microscopy of amniotic fluid cells obtained from normal pregnancies ranging in gestational age from 16 to 22 weeks.

(a) Light microscopy of three large epithelial-like amniotic fluid cells and a small round cell with vacuolated cytoplasm (arrow). One epithelial-like cell demonstrates an indistinct staining nuclei (*) (X180).

(b) Light microscopy of two anucleated amniotic fluid cells (*), a binucleated cell (arrow) and a large epithelial-like cell (X140).

(c) Ultrastructure of a typical nucleated amniotic fluid cell demonstrating an elongated nucleus (Nu) embedded in a matrix of glycogen particles (G). The plasma membrane exhibits numerous microvilli (mv), and a tail-like formation (arrow) (X3200).

(d) Fine structure of another amniotic fluid cell demonstrating a small round nucleus (Nu) embedded in a mixture of glycogen and cellular organelles. The plasma membrane of this cell contains a few microvilli (mv), areas devoid of surface activity (*) and long, finger-like processes (▶) (X4000).

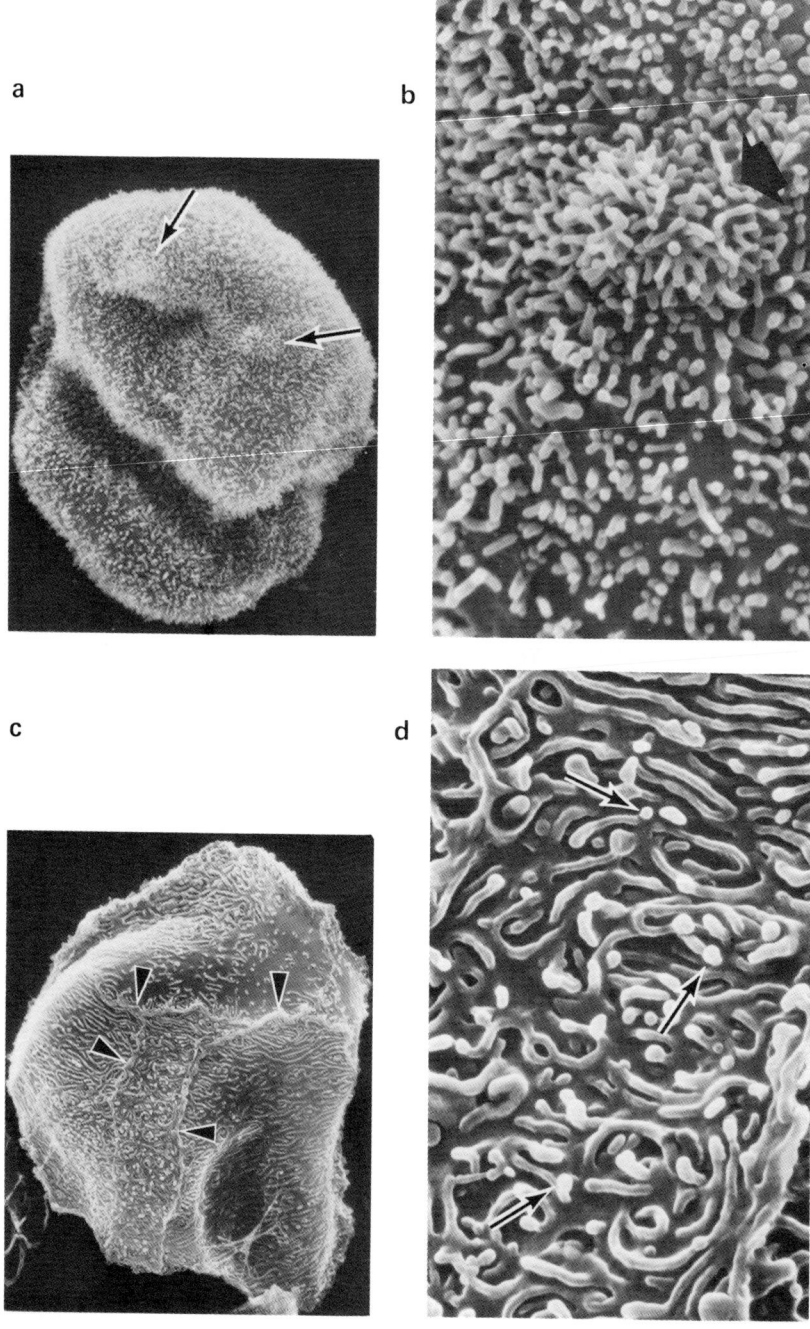

Figure 12.2 Scanning electron microscopy of two nucleated epithelial-like amniotic fluid cells with unique surface topographies.

(a) Surface morphology of a rounded and deeply invaginated cell containing small areas of microvilli concentrations (arrows) (X2100).

(b) High magnification of a portion of the cell surface depicted in Figure 12.2a showing the dense covering of short microvilli (X10,000).

(c) A large amniotic fluid cell characterized by an angular periphery and a plasma membrane containing an extensive network of microplicae formations and intracellular junctions (▶) (X1400).

(d) At higher magnification the microplicae are long and branching with some plicae ending in knob-like formations (arrows) (X10,000).

and scanning electron microscopy (SEM) of normal amniotic fluid cells. The cells presented here were obtained by transabdominal amniocentesis from 15 normal pregnancies ranging in gestational age from 16 to 22 weeks. Upon collection, the cells were immediately fixed in cacodylate-buffered 1% glutaraldehyde and processed for study by TEM or a technique combining LM with SEM (Duff et al., 1976).

LIGHT MICROSCOPY

Light microscopic examination revealed a heterogeneous cytopopulation of amniotic fluid cells. The majority of the cellular populations consisted of epithelial-like nucleated and nonnucleated cells (Figure 12.1a,b). The nucleated cells were usually large with irregular outlines. Occasionally, small round nucleated cells containing vacuolated cytoplasms (Figure 12.1a, arrow) or small binucleated cells were observed (Figure 12.1b, arrow). Anucleated cells or cells containing indistinct staining nuclei or nuclear ghouts were also identified [Figure 12.1a,b, (*)]. Again, these cells varied greatly in shape and size. Anucleate amniotic fluid cells are thought to arise from the various types of nucleated cells by nuclear degeneration (Hoyes, 1968).

There are two principal physiological parameters which influence the cellular population of amniotic fluids: (1) sex of the fetus; and (2) fetal gestation. The proportions of certain cell types vary according to these parameters (Huisjes, 1973). This study is limited to an early stage of gestation in which large epithelial-like cells are the most prevalent.

TRANSMISSION ELECTRON MICROSCOPY

Amniotic fluid cell ultrastructure revealed a mixed cytopopulation of nucleated, nonnucleated and degenerating nucleated cells. A typical large nucleated cell (Figure 12.1c) contained a peripheral ring of cytoplasmic organelles which included rough endoplasmic reticulum, a Golgi complex, mitochondria and lipid inclusions. The remainder of the cytoplasm contained a nucleus of irregular and elongated shape embedded in an even dispersion of glycogen particles. The plasma membrane of this cell demonstrated a dense covering of microvilli of various shapes and lengths and a plasma membrane specialization reminiscent of a tail-like formation (Figure 12.1c, arrow). The fine structure of another nucleated cell (Figure 12.1d) revealed a small nucleus with heavily clumped chromatin surrounded by an intact nuclear membrane, and glycogen granules and cellular organelles that were intermixed and unevenly distributed throughout the cytoplasm. The plasma membrane contained a predominant population of large, long finger-like processes, a few microvilli and membrane areas devoid of surface activity. The intracellular morphology of the amniotic fluid cell populations indicated a variety of functional states and cell types. These two parameters have a direct influence on the culturing of amniotic fluid cells for morphological and/or biochemical diagnoses (Littlefield, 1971; Kaback and Leonard, 1972).

SCANNING ELECTRON MICROSCOPY

SEM of amniotic fluid cells revealed an unexpected and dynamic view of a wide variety of distinctive surface morphologies. All the cells presented here were identified by LM as nucleated, epithelial-like cells prior to SEM examination.

The surface topography of one typical nucleated epithelial-like cell revealed a rounded, deeply invaginated cell (Figure 12.2a) with a plasma membrane exhibiting a dense covering of short microvilli (Figure 12.2b). Several surface areas of this cell demonstrated tufts or concentrations of microvilli formations (Figure 12.2a, arrows). Another larger nucleated amniotic fluid cell revealed an angular-shaped periphery (Figure 12.2c). The plasmalemma of this cell was characterized by an extensive network of microplicae formations (ridge-like folds) and intracellular junctions. The plicae displayed a wide variety of lengths and configurations. Some plicae were long and branching while others appeared to end in knob-like formations (Figure 12.2d, arrows).

Other nucleated amniotic fluid cells revealed a surface containing patterns of microplicae formations separated by elevated,

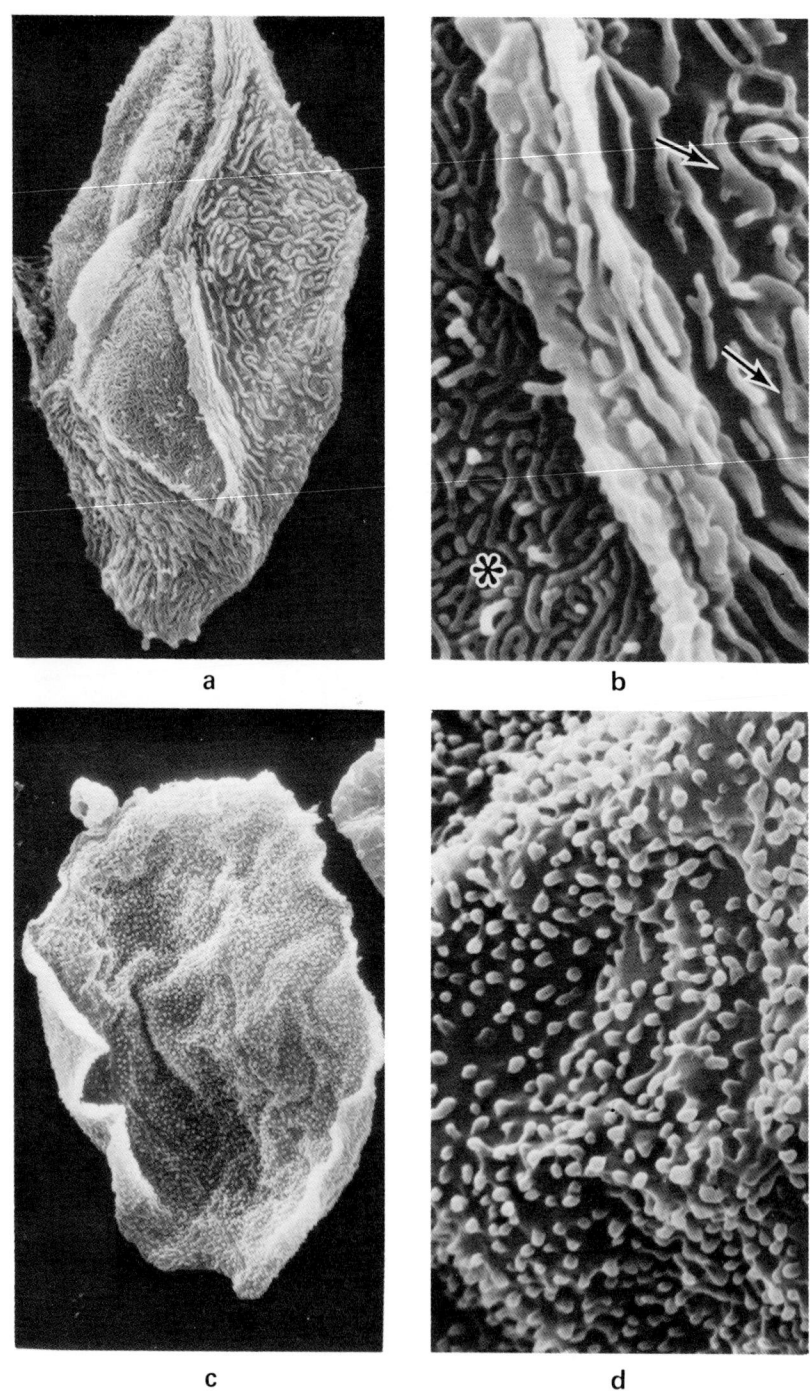

Figure 12.3. Distinctive surface topography of two nucleated amniotic fluid cells.

(a) The surface membrane of an amniotic cell demonstrating two patterns of microplicae formations separated by an elevated intracellular junction (▶) (X2300).

(b) At higher magnification, these two microplicae patterns appear as dense, short branching extensions (*) and thickened, flap-like folds (arrows) (X10,000).

(c) The SEM of an amniotic fluid cell which appears as a thin, floating-cell type (X2100).

(d) At higher magnification, the plasmalemma of this cell exhibits an even distribution of small membrane projections (X10,000).

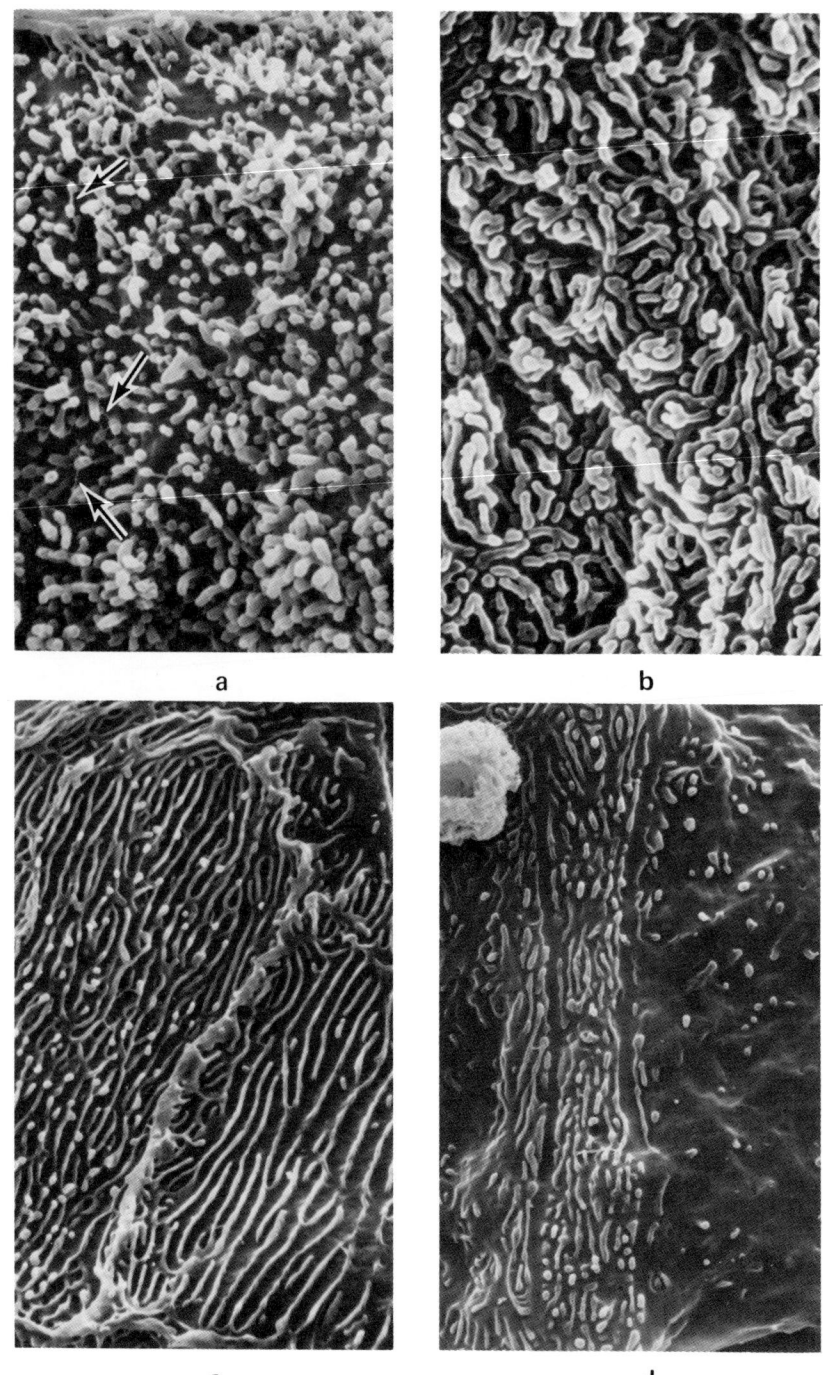

Figure 12.4. Four additional patterns of distinctive surface morphologies observed by SEM on normal amniotic fluid cells.

(a) A plasma membrane surface characterized by microvilli interconnected by small threads (arrows) (X10,000).

(b) An elaborate interdigitating and interconnecting microplicae pattern on another amniotic fluid cell (X10,000).

(c) A plasmalemmae surface consisting of long ridge-like formations that are separated by wide grooves (X5000).

(d) An amniotic cell surface exhibiting a smooth membrane and a limited area of microplicae formation (X5000).

prominent, intracellular borders (Figure 12.3a). One pattern appeared as dense, short-branching plicae with unevenly distributed knob-like or microvilli-like extensions (Figure 12.3b), while microplicae located across the intracellular border appeared as thickened, flap-like folds (Figure 12.3b, arrows).

Surface topography of other nucleated epithelial-like cells exhibited the appearance of thin, floating structures (Figure 12.3c). The plasmalemmae of these cells contained an even distribution of small membrane projections (Figure 12.3d) which are morphologically distinct from microvilli projections of other cells (compare with Figure 12.2b).

Additional unique plasmalemmae patterns were revealed by SEM. Some cells demonstrated a plasma membrane characterized by microvillus projections of various size, shape and distribution (Figure 12.4a) interconnected in some areas by small threads (Figure 12.4a, arrows). Other cells revealed a very elaborate interdigitating and interconnecting pattern of microplicae formations (Figure 12.4b). By contrast, other cells demonstrated a plasmalemmal surface of long ridge-like formations separated by wide grooves (Figure 12.4c) while other cells presented a relatively smooth plasmalemma with only occasional small projections and defined areas of microplicae formations (Figure 12.4d). The unique surface features of amniotic fluid cells indicate a wide variety of cell types with presumably different functional capabilities. Microplicae surface formations similar to some amniotic cells (Figures 12.2c, 12.3a and 12.4b,c) have been identified in animal systems studied by SEM (Andrews, 1976) as characteristic of surfaces consisting of stratified squamous epithelial cells (oral mucosa, esophagus, anal canal, tongue). Thus, this correlation suggests the potential usefulness of SEM identification of the specific types and sources of amniotic fluid cells.

CONCLUDING REMARKS

This study was undertaken in an attempt to increase our basic knowledge concerning the morphological parameters of normal amniotic fluid cells obtained during the gestational period used for *in utero* detection of hereditary abnormalities. Hopefully this knowledge will contribute information towards the questions of origin and function of normal cells, and provide a basis for the future examination of abnormal amniotic fluid cells.

ACKNOWLEDGMENTS

The authors wish to thank Mrs. Bridget Duff and Mrs. Juanita Clark for their excellent technical assistance throughout the completion of this project.

REFERENCES

Andrews, P. M. "Microplicae: Characteristic Ridge-like Folds of the Plasmalemma," *J. Cell Biol.* 68:420 (1976).

Bourgeois, G. A. "The Identification of Fetal Squamas and the Diagnosis of Ruptured Membranes by Vaginal Smear," *Am. J. Obstet. Gynecol.* 44:80 (1942).

Duff, B. A., S. M. Noonan and L. Weiss. "The Cytology and Surface Topography of Normal Amniotic Fluid Cells," *Micron* 7:299 (1976).

Fuchs, F. and L. L. Cederquist. "Prenatal Diagnosis Based Upon Amniotic Fluid Cells," in *Amniotic Fluid*, D. V. I. Fairweather and T. K. A. B. Eskes, Eds. (Amsterdam: Excerpta Medica, 1973), pp. 262-276.

Hoyes, A. D. "Ultrastructure of the Cells of the Amniotic Fluid," *J. Obstet. Gynaecol. Brit. Commonwlth.* 75:164 (1968).

Huisjes, H. J. and J. H. Arendzen. "Estimation of Fetal Maturity by Cytologic Evaluation of Liquor Amnii," *Obstet. Gynecol.* 35:725 (1970).

Huisjes, H. J. "Cytology of the Amniotic Fluid and its Clinical Applications," in *Amniotic Fluid*, D. V. I. Fairweather and T. K. A. B. Eskes, Eds. (Amsterdam: Excerpta Medica, 1973), pp. 95-132.

Jacobson, C. B. and R. H. Barter. "Intrauterine Diagnosis and Management of Genetic Defects," *Am. J. Obstet. Gynecol.* 99:796 (1967).

Kaback, M. M. and C. O. Leonard. "Morphological and Enzymological Considerations in Antenatal Diagnosis," in *Antenatal Diagnosis*, A. Dorfman, Ed. (Chicago: University of Chicago Press, 1972), pp. 81-94.

Littlefield, J. W. "Problems in the Use of Cultured Amniotic Fluid Cells for Biochemical Diagnoses," *Birth Defects* 7:15 (1971).

Nadler, H. L. and A. Gerbie. "Present Status of Amniocentesis in Intrauterine Diagnosis of Genetic Defects," *Obstet. Gynecol.* 38:789 (1971).

SELECTED REFERENCES

Barnhart, M. I. and B. Mandelbaum. "Amniotic Fluid Cells," in *SEM Atlas of Mammalian Reproduction*, E. S. E. Hafez, Ed. (Tokyo: Igaku Shoin Ltd., 1975), pp. 376-385.

Bourne, G. *The Human Amnion and Chorion* (Chicago: Year Book Medical Publ., Inc., 1962).

Hafez, E. S. E, M. I Barnhart, H. Ludwig, J. Lusher, I. Joelsson, J. L. Daniel, A. T. Sherman, J. A. Jordan, H. Wolf, W. C. Stewart and F. C. Chrétien. "Scanning Electron Microscopy of Human Reproductive Physiology," *Acta Obstet. Gynecol. Scand.* Suppl. 40:1-61 (1975).

SECTION IV

PRODUCTS OF CONCEPTION

CHAPTER 13

THE MAMMARY GLANDS

Elinor Spring-Mills and Joel J. Elias

GENERAL STRUCTURE AND FUNCTION

The mammary glands are specialized accessory skin glands located in the subcutaneous tissue (superficial fascia). In adult women, each breast contains 15-20 lobes radiating from the nipple. An individual lobe is surrounded by broad bands of connective tissue and is drained by an independent excretory or lactiferous duct which terminates in the nipple. In addition, each lobe is subdivided into numerous lobules or gland fields by layers of connective tissue rich in adipocytes.

The lobule is the basic structural unit of the breast. Depending upon the age and physiological status of the woman, each lobule is composed of 10-100 alveoli which empty into a common intralobular duct. Ducts proliferate during pregnancy and new alveoli grow out from the terminal ductules. Full structural and functional maturation is not complete, however, until after childbirth, during lactation, when the terminal ductules and alveoli synthesize and secrete milk. The gland undergoes involution following lactation and atrophic alterations after the menopause.

Unless otherwise indicated in the figure legends, all scanning electron micrographs in this chapter show inactive or resting adult mammary gland tissue which was obtained from women undergoing reduction mammoplasty. These specimens are free of overt disease and, theoretically at least, should more accurately illustrate breast morphology than so-called normal regions from breasts with benign or malignant neoplasms (Spring-Mills and Elias, 1975).

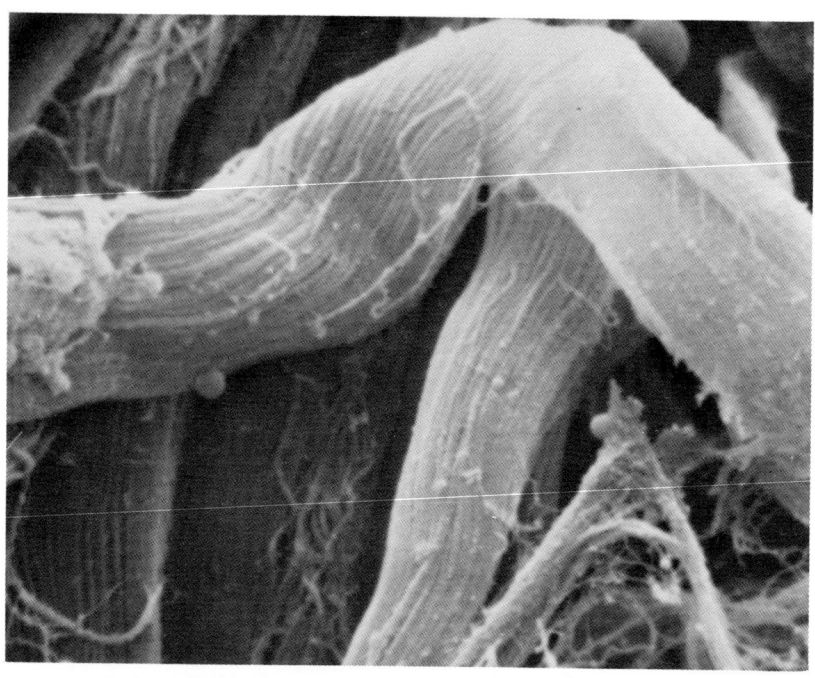

Figure 13.1. The collagen fibers in the interlobular connective tissue are much thicker than the collagen fibers within the lobules. The thick fibers are composed of many thin fibrils (X11,200).

Figure 13.2 These collagen fibers are smaller in diameter than those shown in Figure 13.1, but form a fairly close meshwork within the stroma of the gland. In fibrous disease of the breast there is a selective increase in the amount of fibrous stroma and a reduction in the number of lobules. In other breast diseases, the epithelium and stroma are both stimulated, although often unequally (X6200).

RESTING GLAND: STROMA

The interlobular connective tissue is quite dense (Figures 13.1 and 13.2), while the intralobular connective tissue is more cellular and contains less fat. The loose, irregular connective tissue surrounding the intralobular ducts allows for distensibility and proliferation of the parenchyma.

RESTING GLAND: PARENCHYMA

Duct and Alveolar Epithelium

Few true alveoli with simple cuboidal epithelium are present in the resting breast. Most of the ducts contain bilayered, stratified cuboidal or columnar epithelium (Figures 13.3 and 13.4). The apical surfaces of the epithelial cells are covered by microvilli and at least two types of epithelial cells can be found bordering the lumens of the secondary ducts and terminal ductules (Figures 13.5 and 13.6).

Myoepithelium

A discontinuous layer of contractile myoepithelial cells can be found between the epithelial cells and the basal lamina of the ducts and alveoli. They are elongated and stellate in shape (Figure 13.7). They are arranged in spirals around the ducts and alevoli and presumably provide a contractile mechanism for ejecting milk from the lumens.

ACKNOWLEDGMENTS

The authors wish to thank Ms. Maria Maglio for specimen preparation and Mrs. Naomi Ross for secretarial assistance.

Products of Conception 197

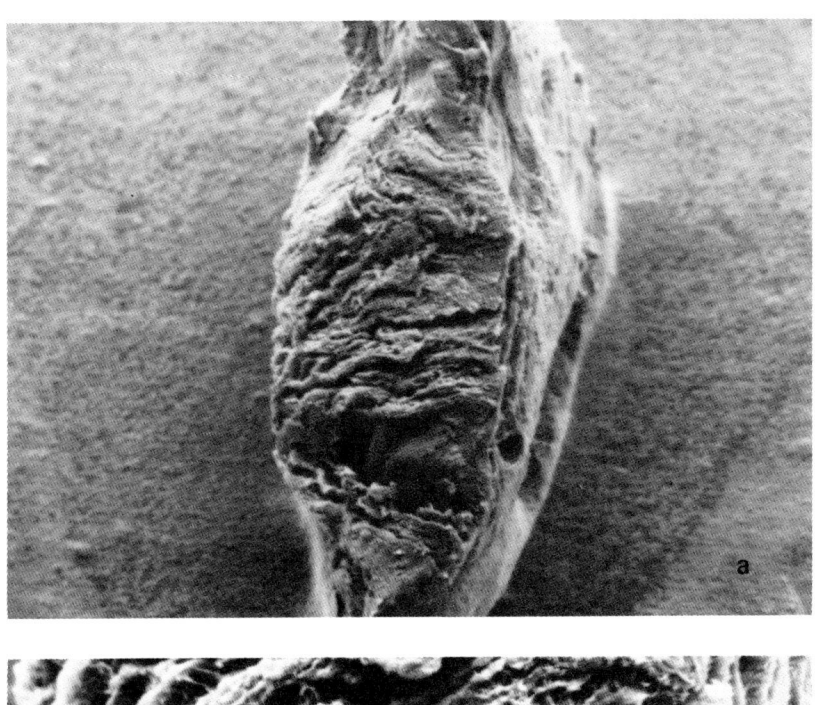

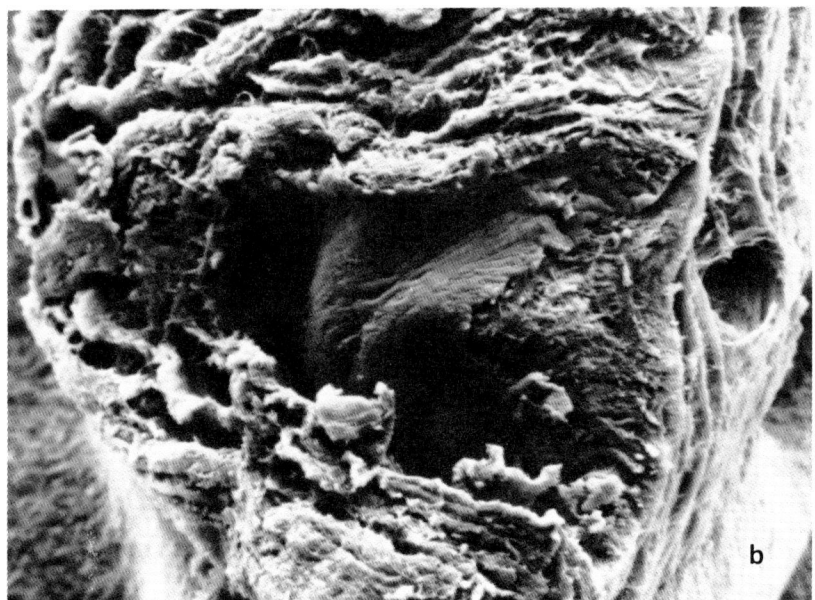

Figure 13.3. A large duct surrounded by dense connective tissue is shown here at two different magnifications. The wall of the duct has some longitudinal folds but is free of the papillomatous stalks and projections present in hyperplastic ducts of patients with benign and malignant breast diseases: (a) X63; (b) X176.

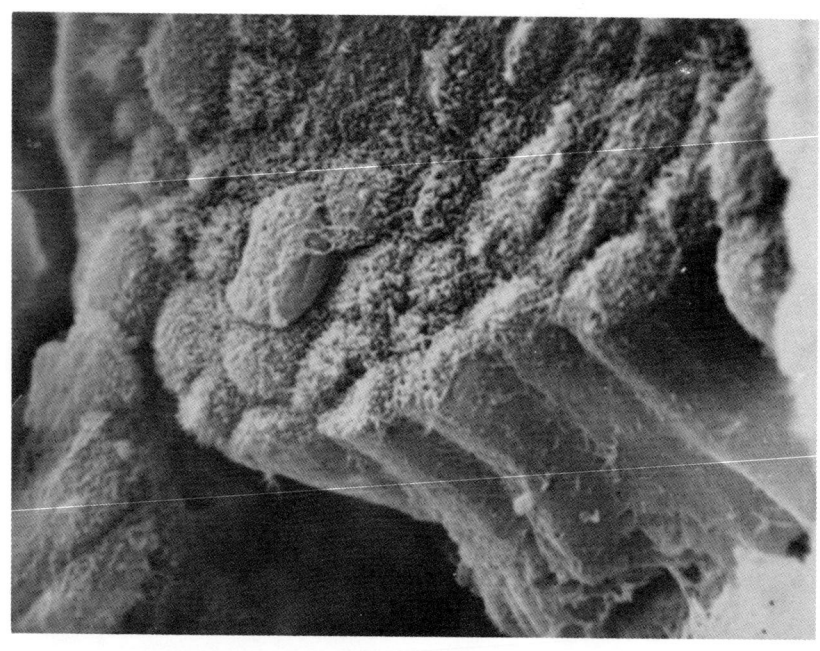
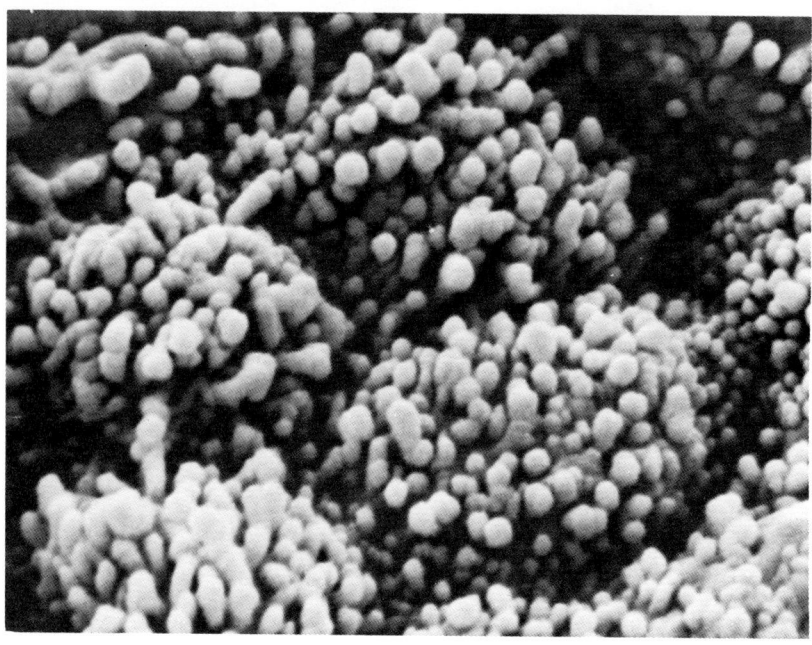

Figure 13.4. The apical and lateral surfaces of columnar duct cells from a patient with infiltrating duct carcinoma are shown in this scanning electron micrograph. Microvilli are present on the apical surfaces and applications which presumably facilitate cell adhesion are present on the lateral cell membranes (X3155).

Figure 13.5. Most duct and ductule cells resemble the cells shown in this scanning electron micrograph. Their microvilli are short and thick (X16,000).

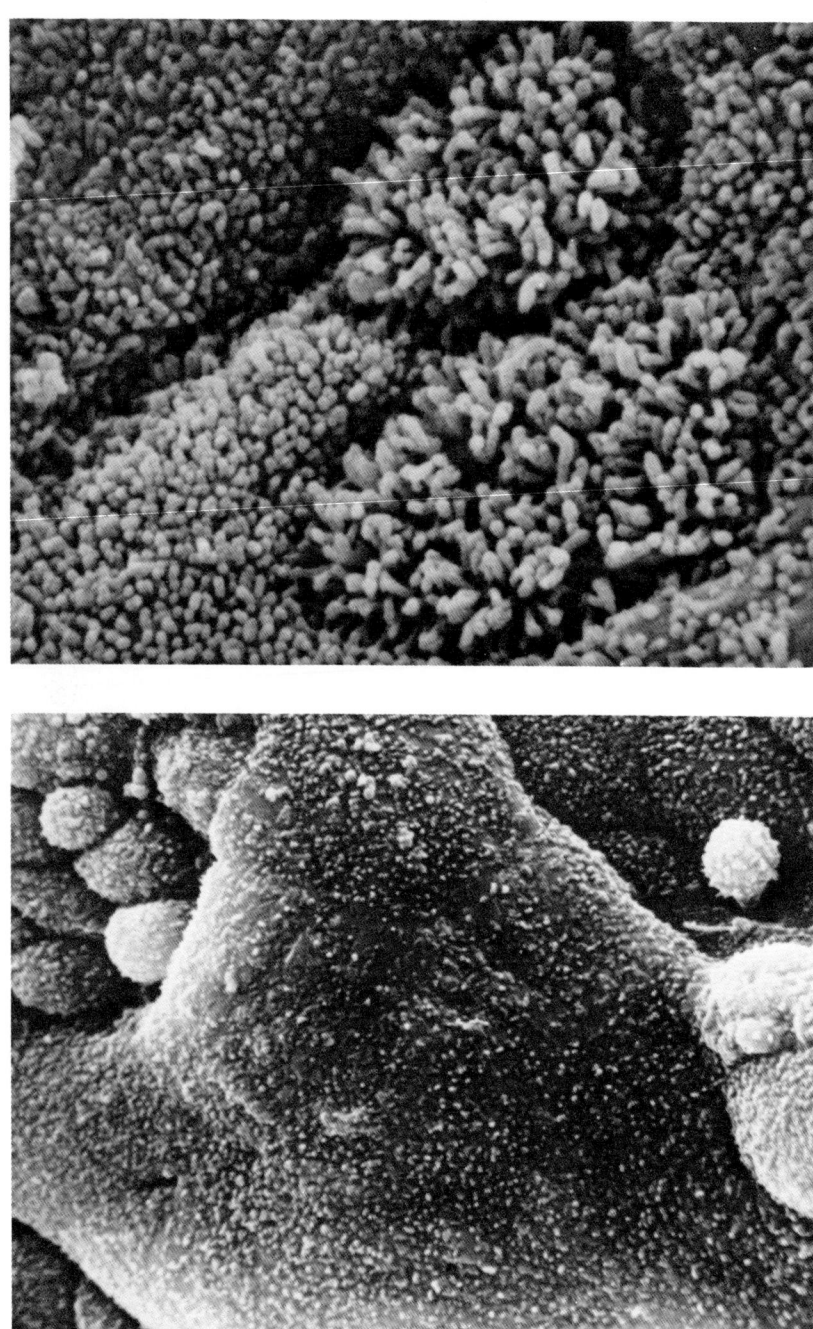

Figure 13.6. Cells with longer microvilli often are scattered at intervals along the ducts. Additional research is required, however, to define the differences between these cells and those with the shorter microvilli (X13,600).

Figure 13.7. A portion of the surface overlying the cell body of a myoepithelial cell is shown in this biopsy specimen from a woman with fibrocystic disease (X2933).

REFERENCES

Bargmann, W. and A. Knoop. "Über die Morphologie der Milchsekretion. Licht-und elektronenmikroskopische Studien an der Milchdrüse der Ratte," *Zeitschr Zellforsch.* 49:344 (1959).

Cowie, A. T. and S. J. Folley. "The Mammary Gland and Lactation," in *Sex and Internal Secretions*, W. C. Young, Ed. (Baltimore: Williams and Wilkins Co., 1961), p. 590.

Foote, F. W. and F. W. Stewart. "Comparative Studies of Cancerous versus Non-Cancerous Breasts. I. Basic Morphological Characteristics," *Ann Surg.* 121:6 (1945).

Haagensen, D. D. *Diseases of the Breast* (Philadelphia: W. B. Saunders Co., 1971).

Kon, S. K. and A. T. Cowie. *Milk: The Mammary Glands and Its Secretion* (New York: Academic Press, 1961).

McDivitt, R. W., F. W. Stewart and J. W. Berg. *Atlas of Tumor Pathology: Tumors of the Breast* (Washington, D.C.: Armed Forces Institute of Pathology, 1968).

Spring-Mills, E. and Y. J. Topper. "Some Ultrastructural Effects of Insulin, Hydrocortisone and Prolactin on Mammary Gland Explants," *J. Cell Biol.* 44:310 (1970).

Richardson, K. C. "Contractile Tissues in the Mammary Gland, with Special Reference to the Myoepithelium in the Goat," *Proc. Roy. Soc. B.* 136:30 (1949).

Spring-Mills, E. and J. J. Elias. "Cell Surface Differences in Ducts from Cancerous and Noncancerous Human Breasts," *Science* 188:947 (1975).

Spring-Mills, E. and J. J. Elias. "Cell Surface Changes Associated with Human Breast Cancer," *Scanning Electron Microscopy/1976 Proc.* II:1 (1976).

Wellings, S. R., K. B. DeOme and D. R. Pitelka. "Electron Microscopy of Milk Secretion in the Mammary Gland of the $C_3H/Crgl$ Mouse. I. Cytomorphology of the Prelactating and Lactating Gland," *J. Natl. Cancer Inst.* 25:393 (1960).

Wellings, S. R., B. W. Grunbaum and K. B. DeOme. "Electron Microscopy of Milk Secretion in the Mammary Gland of the $C_3H/Crgl$ Mouse. II. Identification of Fat and Protein Particles in Milk and in Tissue," *J. Natl. Cancer Inst.* 25:423 (1960).

CHAPTER 14

THE EMBRYO AND FETUS

J. H. L. Watson and M. A. Kamash

There are two major periods in the development of the fetus: (1) during the fourth to the eight weeks, known as the embryonic period, in which the three germ layers (embryonic ectoderm, mesoderm and endoderm) give rise to a number of specific tissues and organs; and (2) from the beginning of the third month to the end of intrauterine life which is known as the fetal period and is characterized by maturation of the tissues and rapid growth of the body (Langman, 1975).

In general, during the embryonic period the main germ layer derivatives are: (1) the ectoderm, giving rise to the central nervous system, the peripheral nervous system, the sensory epithelia of the eye, ear and nose, the epidermis and its appendages (hair and nails), the mammary glands, the hypophysis, subcutaneous glands, and the enamel of teeth; (2) the mesoderm, giving rise to cartilage bone and connective tissue, stratified and smooth muscles, the heart, blood and lymph vessels and cells, the kidneys, the gonads and genital ducts, the serous membranes lining the body cavities, the spleen, and the cortex of the suprarenal gland; and (3) the endoderm, from which originates the epithelial lining of the gastrointestinal and respiratory tracts, the parenchyma of the tonsils, thyroid, parathyroids, thymus, liver and pancreas, the epithelial lining of the urinary bladder and urethra, and the epithelial lining of the tympanic cavity, tympanic antrum and auditory tube (Moore, 1977).

On the other hand, the fetal period begins about nine weeks after fertilization and ends at birth. Lanugo, head hair and eyebrows appear during the third month. The limbs reach their relative length in comparison to the rest of the body, and the external genitalia develop. The fetus lengthens rapidly and several organ systems are able to function by the sixth month or in the first half of the seventh. By the end of intrauterine life the skin is covered by the vernix caseosa, a whitish, fatty substance which is composed of the secretory products of the sebaceous glands (Langman, 1975).

There is a slowing of growth as the time of birth approaches. Fully developed fetuses usually reach a crown-rump length of 360 mm and weigh about 3,000-3,500 g at term. In general, male fetuses grow faster than females, and male infants generally weigh more than females at birth.

Scanning electron microscopy (SEM) is a powerful tool for examining the surface architecture of the embryo and fetus and for revealing the topographical relationship between surface cells. This chapter describes from SEM the details of early outer surface structure of both a normal human embryo (41 days) and fetus (61 days) and correlates these with current knowledge derived from light microscopy. It also describes briefly the SEM of abnormal development of a conceptus resulting from spontaneous abortion.

SIX-WEEK-OLD EMBRYO

Light microscopy shows that during the fifth and sixth weeks of development, organogenesis continues on a wide front with the following structures appearing for the first time: cerebral hemispheres, cerebellum, olfactory pits, primitive retina, parathyroids, thymus, primary, secondary and tertiary bronchi, ventral pancreatic diverticulum, spleen, cardiac septa, ureteric buds, gonadal ridges, and genital tubercle. Also, during this week the stomach begins to rotate and the midgut forms a loop. Externally the face is forming and the limb buds exhibit their basic plan of limb, forelimb and hand or foot. By the sixth week, suprarenal glands begin to form, the secondary palate appears, the venae cava take shape histologically, and the gonads become identifiable as either testes or ovaries. The finger rays, wrist and elbow points can be recognized. In addition, pigment appears in the retina (Patten, 1947).

The 41-day-old embryo observed by SEM (Watson et al., 1976a) from the left lateral surface (Figure 14.1a) is characterized by development of the skull at the telencephalic and mesencephalic protuberances, the corneal ectoderm with the upper and lower eyefolds, the nasal bridge, the external auditory meatus and its developing cartilages and the mandible developed from the bronchial arch. The cardiac and hepatic prominences are also noted. The upper limb is well-formed and the first suggestions of digital budding are observed. The lower limb is also well-differentiated although, as expected, the digits are less evident than those of the arm.

The forming forearm and foreleg are delineated, as are the curved caudal end of the lower spine and the umbilicus. The surface of the ectoderm along the forming spine is wrinkled by the processing for SEM, as it is also over the top and back of the head.

At a dorsal angle (Figure 14.1b), the two lower limbs are shown and a suggestion of phalangeal development is noted in the right foot pad. The wrinkled nature of the ectoderm over the spine and tail is again evident and the umbilicus is recognized. In a ventral view, the tail to the left and the conical elevation of the genital tubercle to the right are seen (Figure 14.1c). On the caudal aspect of the tubercle are the paired genital folds flanking a longitudinal depression (urogenital sinus), which extends toward the caudal extremity and in which there is a suggestion of two openings. On either side of the genital folds and farther out laterally are the paired, vaguely visible elevations of the labioscrotal swellings.

In the head, nasomedial processes fuse with the maxillary processes to form the upper lip (Figure 14.1d). The nasal pits, nasolateral processes and other structures seem to be similar to those expected by light microscopy at seven weeks. Within the area of the corneal ectoderm is a rounded area (arrow), diameter about 0.02 cm, which probably represents the surface of the corneal ectoderm stretched over the lens forming beneath it.

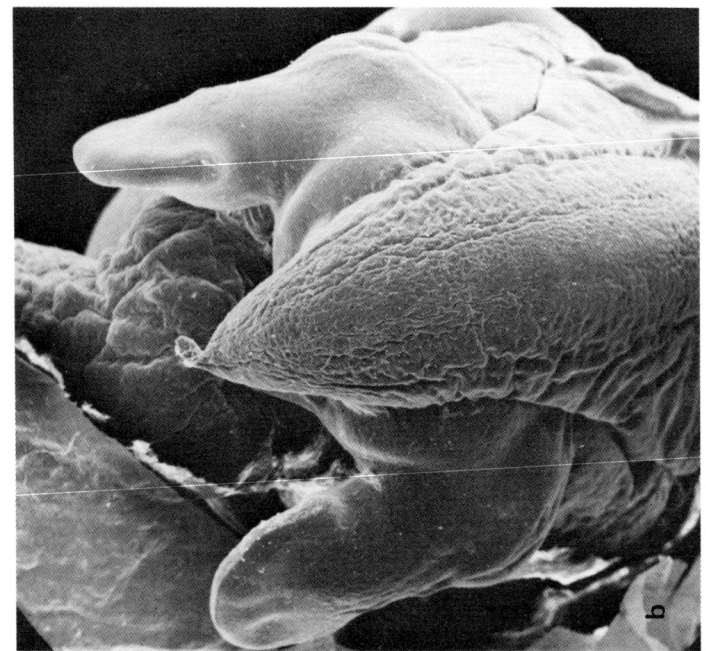
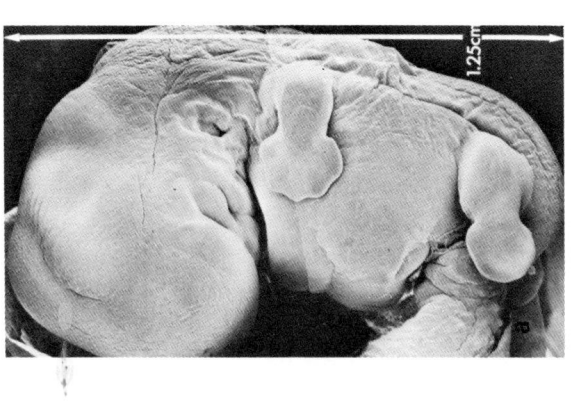

Figure 14.1. (a) Human embryo at 41 days by SEM. (b) Dorsal view of the lower extremities of the human embryo at 41 days (X13).

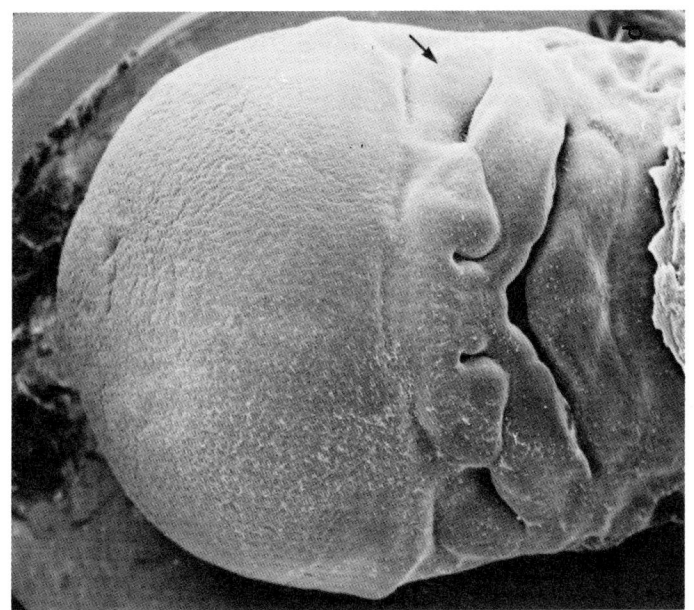
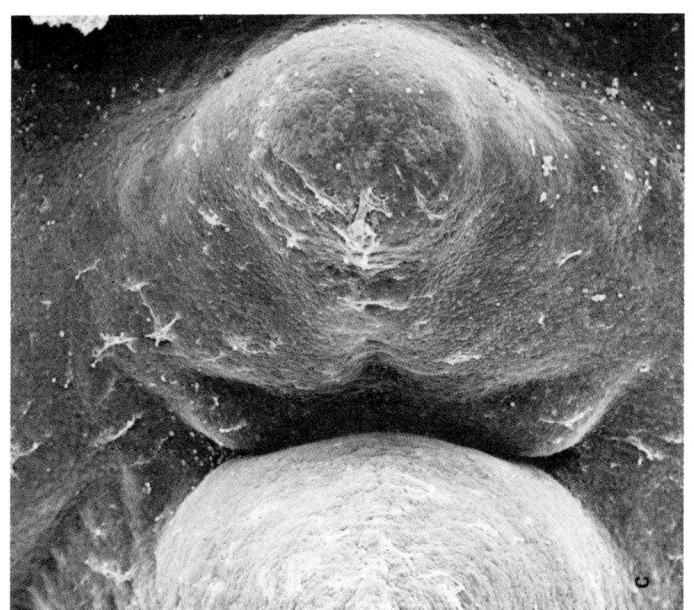

(c) Ventral view of the tail (left) and the sex tubercle (right) of the human embryo at 41 days (X60).
(d) Face of the human embryo at age 41 days (X18).

THE ECTODERM OF THE 41-DAY-OLD EMBRYO

SEM of the cells of the surface of the ectoderm show similarities at the following locations: the ear, dorsal and ventral tips of the forming digits of the upper limbs, dorsum and palm of the hands, top of the head, forearm, cardiac and hepatic prominences, sides of the tail and of the lower limbs. It is therefore concluded that there is an identity in the surface of the ectoderm everywhere over the surface, except for minor local variations. The cells tend to be closely packed in sheets, often with raised, well-defined boundaries (Figure 14.2a). They are slightly raised over the region of their nucleus, but there are no rounded cells. There is relatively little surface activity, except that manifest by a sparse population of stubby microvilli, which are mostly evenly distributed over the cell surfaces. Such a pattern in the microvillous population involves a tendency for microvilli to be concentrated on some cells at cell centers and/or at the margins in some locations (Figure 14.2a). Such minimal surface activity is noted over the hepatic and cardiac prominences. A few smooth-surfaced cells with few or no surface projections are found to be interspersed among the microvillous cells.

Only the cells of the corneal ectoderm show obviously different characteristics of surface structure at 41 days (Figure 14.2b). While the ectoderm elsewhere is comprised of sheets of fairly flat microvillous cells as described, that over the cornea is comprised of markedly raised or cobblestone cells arranged in an almost linear pattern. Cells of the corneal ectoderm have few if any microvilli and are somewhat smaller than the flat cells at other locations.

NINE-WEEK-OLD FETUS

As the embryonal period approaches completion, only the respiratory and genital systems, neither of which needs to be functional *in utero*, appear to be markedly undeveloped. The rest of the systems are easily recognizable and several of them, particularly the cardiovascular, are already functional to some degree. Differentiation ceases to be a major factor and the remainder of prenatal development becomes primarily a function of growth.

The fetus is about 2.5 cm long at eight or nine weeks and by light microscopy, all of the internal organs as well as the bones, muscles, nerves and major vessels can be identified. At this time, the eyelids are forming and the nose, jaws, external ears, fingers and toes are distinct. The elongation of the neck has begun to separate head from thorax, but the coils of the gut still lie in the colon of the umbilical cord.

Through SEM, the head of the 61-day-old fetus is again collapsed and crumpled due to difficulties in processing the specimen for examination, but in spite of this it is possible now to identify the normal formed facial features (Watson *et al.*, 1976a,b). Both the upper and lower limbs, the hands and feet with their digits and phalanges are well differentiated and the flesh pads are recognized (Figure 14.2c).

The ectoderm at 61 days is almost everywhere composed of cells which are fairly flat, well raised over the nucleus and heavily microvillous. The microvilli are heavily populated over the entire cell surface but with concentrations toward the centers over the nucleus and at cell margins. Most of the microvilli are long and well formed. In addition, the cells now demonstrate frequent intercellular bridging (Figure 14.2d). A few isolated cells are heavily blebbed. Although the digits and phalanges are well differentiated, there are still no surface differences between the ectoderm of their dorsal and ventral sides in the region of the future nails, nor any evidence that the nails are forming.

Local variations in cell types are observed to occur toward the ventral and dorsal tips of the phalanges. At their extreme tips all cells are rounded and raised in a cobblestone surface. Away from the third phalange flatter cells begin to predominate and at the second phalange all the cells are flat cells in sheets where the cells have little, if any, defined, visible boundaries or margins. They are liberally supplied with long microvilli and cytoplasmic bridges interconnecting neighboring cells, giving the surfaces an appearance similar to cells in flat culture. As measured by the variety and density of their surface specializations, the cells of the ectoderm at 61 days are very active everywhere.

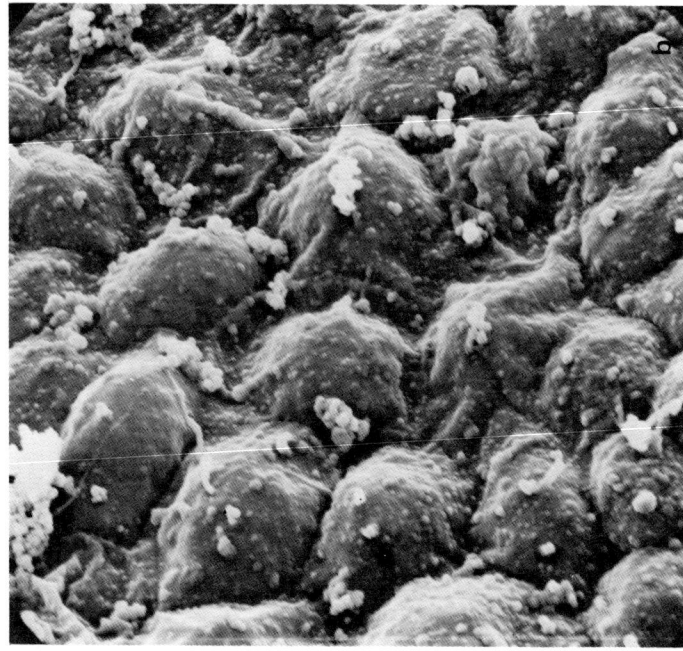
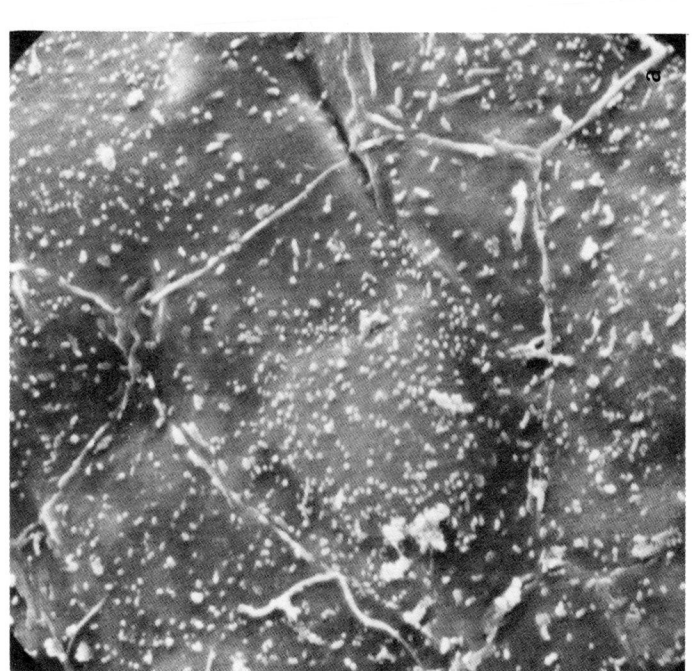

Figure 14.2. (a) Cellular surface of ectoderm on the side of the right limb of human embryo at 41 days shows concentration of microvilli over the nucleus (X4000). (b) Cellular surface of corneal ectoderm of human embryo at 41 days (X2600).

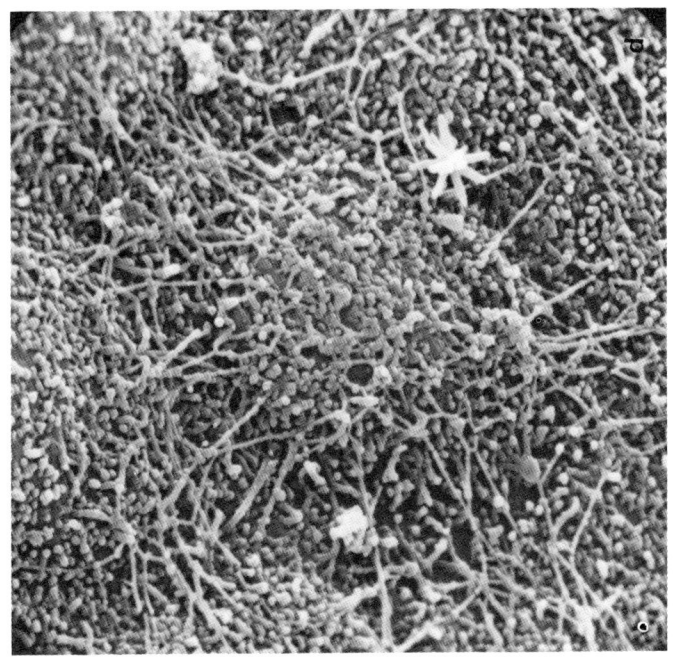

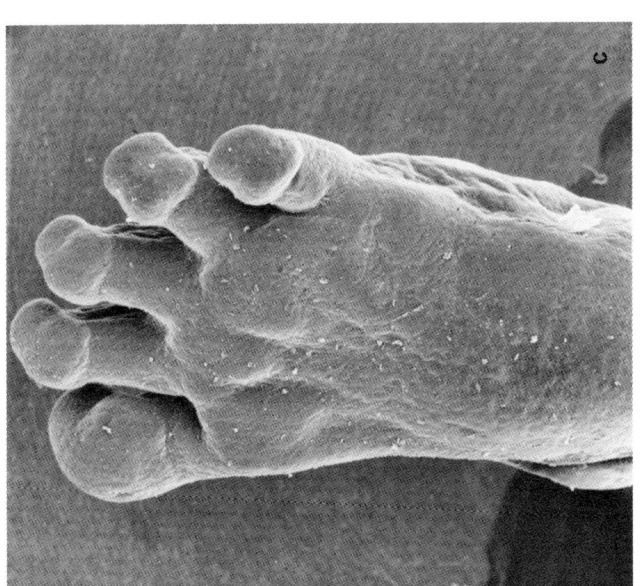

(c) Well-differentiated left foot of 61-day-old fetus showing the phalanges and flesh pads (X21).
(d) Ectoderm over the ventral surface of the first pedal digit of a 61-day-old fetus showing sheets of slightly raised microvillous cells with intracellular bridges (X7000).

Figures 14.3a,b show flat cells of the ectoderm located relatively close to the growing tips of a developing third phalange. The arrow in each lies in the direction of phalange growth. It is seen that the margins of the cells and the intercellular bridges stretch out parallel to the direction of growth but tend to bunch up at right angles to it, so that they become raised abruptly and covered with a mass of intertwining microvilli and intercellular bridges from the adjoining cells. Distal from the growing region of a digit, as in Figure 14.2d, there is uniformity in the appearance of the cell margins around the cell regardless of direction.

COMPARISON OF THE 41-DAY-OLD EMBRYO WITH THE 61-DAY-OLD FETUS ECTODERM

As observed by SEM, the ectodermal (peridermal) cells at 61 days are quite different from those of the ectoderm at 41 days. The relatively low surface activity, represented by the sparsely microvillous surface of the sheets of cells at 41 days has yielded to sheets of cells at 61 days which are heavily microvillous and are even occasionally blebbed. Blebbing is never observed at 41 days. No ruffling of cell margins is observed in the ectoderm surface of either the embryo or the fetus.

The microvilli at 61 days are longer and better formed than the stubby projections observed at 41 days, and appear to be true microvilli. Their presence increases considerably the active cellular surface of the ectoderm in the 61-day-old fetus over that of the 41-day-old embryo, and thereby offers increased surface exposure to the amniotic fluid.

While there are no observed cobblestone areas at 41 days, such raised or rounded cells are observed at 61 days, usually on the growing tips of the phalanges. However, even the sheets of cells seem to be somewhat more raised over the nuclei at 61 than at 45 days.

No intercellular bridging is observed at 41 days but at 61 days there are many examples where the intercellular "strands" of cytoplasm are observed to originate from one and end at another cell's surface. However, there are two general types of such strands at 61 days. The wider ones are about 100-150 nm wide and occur over cell boundaries to obscure them. These are interpreted here as "intercellular bridges," connecting cell to cell. Finer, beaded fibers, which are from 5-80 nm wide, frequently project directly from the cell surface, or form networks well above it. These may be either mucus or fibrin or represent acid mucopolysaccharides (Hayes, 1967) known to be present on embryo surfaces.

ABNORMAL DEVELOPMENT

Spontaneous abortion is defined as the termination of pregnancy before 20 weeks gestation. The reasons for lack of success of an implanted pregnancy can be grouped in several categories, including genetic error in the embryo which is common among abortuses and/or structural deficiencies in the uterus.

The following is a description of human abortus examined by SEM. The specimen is from a spontaneous abortion at 78 days following conception date (Watson *et al.*, 1976a), in which the stunted and disorganized remains of a human embryo appear to have developed to about the fourth week. Only the protuberance of the forebrain, the mandibular arch and the possible depression of the auditory pit are recognizable in the face. In the body, the cardiac prominence is identifiable, along with the head swelling and the tail. No limb buds are present and there is no open neural fold.

The surface of the ectoderm is quite smooth over the tail but crumpled elsewhere, particularly over the dorsal areas. The squamous cells of the ectoderm resemble those of the 41-day-old embryo, having little surface structure except for tiny blebbing and no obvious microvilli. Many ectodermal areas show signs of advanced autolysis and deterioration. There are many polymorphonuclear leukocytes and lymphocytes on the surfaces, particularly on the head and at the edge of the mandibular arch, often under a heavy, fibrous coating. Light microscopy reveals nothing to identify the specimen as an embryo or fetus, except some suggestion of kidney tubule formation.

An important relationship is known to exist between abnormal placenta and the occurrence of spontaneous abortion. The

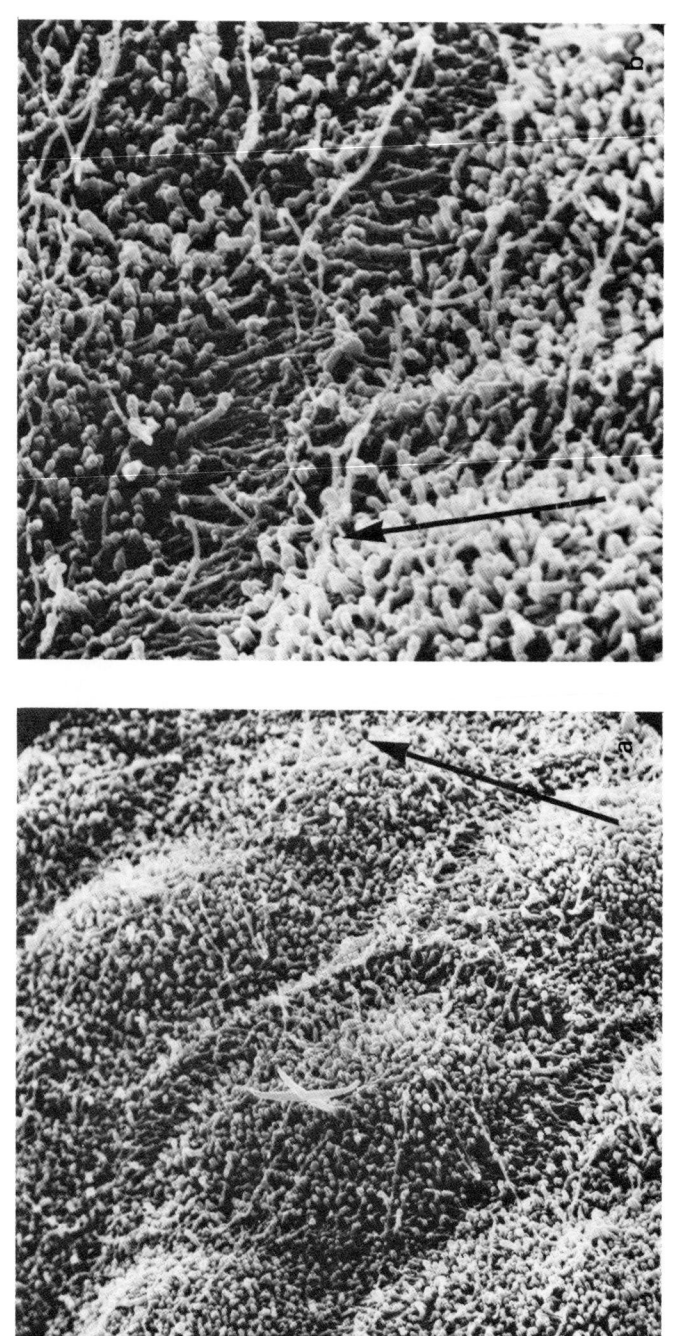

Figure 14.3. (a) Ectoderm close to the growing tip of a developing phalange, demonstrating stretching of the cell in the direction of growth (arrow) (X5000). (b) Same ectoderm as in Figure 14.3a at higher magnification. The arrow in the direction of growth (X10,000).

Products of Conception 215

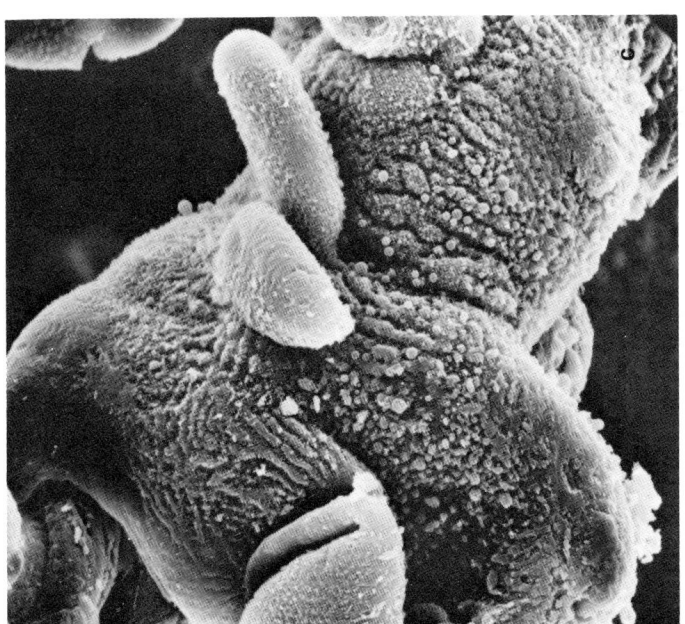

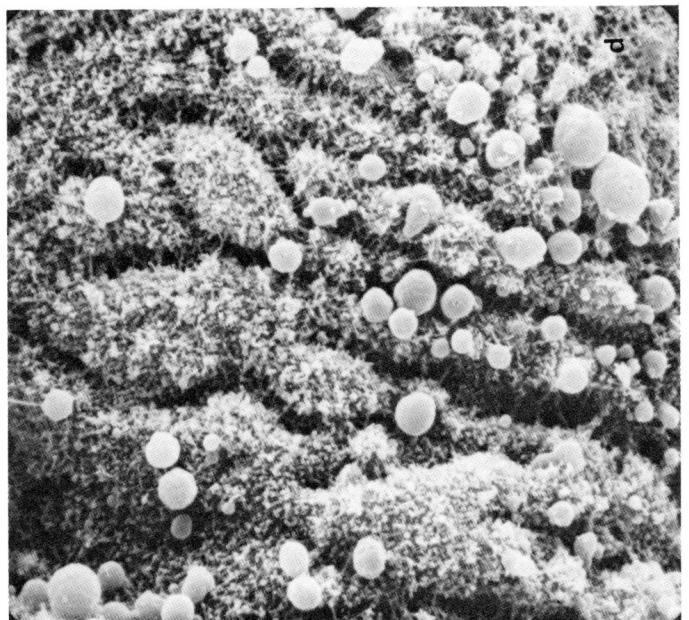

(c) Branched chorion of spontaneously aborted conceptus (X400).
(d) Surface of villus of the chorion of spontaneously aborted conceptus showing microcysts (X2000).

villi of the chorion (Figure 14.3c) from the same conceptus are not grossly swollen and show normal branching. In many regions (Figures 14.3c,d), the villi are cystic, bearing smooth-surfaced cysts with diameters ranging from 220nm-1μ, and resembling in these microsizes the hydatidiform degeneration reported for triploidy (Carr, 1971). It is at least as probable, however, that these cystic structures can be a product of the autolytic changes. The finest ultrastructure (Figure 14.3d) of the villi surface is the mass of long microvilli (about 130 nm in diameter) observed in the background of Figure 14.3d, a structure which lends the necessary high specific surface to the villi of the chorion.

REFERENCES

Carr, D. H. *Advances in Human Genetics,* H. Harris and K. Hirschhorn, Eds. (New York: Plenum Press, 1971), Chapter 4, pp. 201-257.

Hayes, A. D. "Acid Mucopolysaccharide in Human Fetal Epidermis," *J. Invest. Derm.* 48:598 (1967).

Langman, J. *Medical Embryology*, 3rd ed. (Baltimore: Williams & Wilkins, 1975).

Moore, K. L. *The Developing Human*, 2nd ed. (Philadelphia: W. B. Saunders, 1977).

Patten, B. M. *Human Embryology* (Philadelphia: The Blakeston Co., 1947), p. 429.

Ried, D., K. L. Ryan and K. Benirschke. *Principles and Management of Human Reproduction* (Philadelphia: W. B. Saunders, 1972).

Watson, J. H. L., A. Asfari, B. H. Drukker and J. L. Swedo. *IITRI/SEM* VI:393-402 (1976a).

Watson, J. H. L. and J. L. Swedo. *J. Microscop. Soc. Can.* III:140 (1976b).

Watson, J. H. L., J. L. Swedo and B. H. Drukker. *Proc. 34th Ann. Electron Microscopy Soc. Am.*, Miami Beach, Florida, 152 (1976c).

CHAPTER 15

UMBILICAL CORD & NEONATAL BLOOD

Marion I. Barnhart and Jeanne M. Lusher

UMBILICAL CORD

The umbilical cord, which is of fetal origin, contains two arteries and one vein which follow a spiral course to give the cord its twisted rope appearance. These blood vessels alone lack vasa vasorum so that nourishment and waste elimination depend on diffusion to and from the circulating blood. The umbilical vessels are surrounded by an abundant connective tissue unusually rich in mucopolysaccharide (Wharton's jelly). By SEM this mucous connective tissue appears as interlacing nets of collagen with occasional fibroblasts and macrophages (Figure 15.1). The wide spaces between collagen bundles are where Wharton's jelly was present prior to some extraction by the SEM preparative technique. The umbilical cord has epithelial layer or layers depending on gestational age and is further encased by amnion.

Umbilical vessels are relatively thick-walled and muscular (Figures 15.2 and 15.3). Frequently they are contracted along much of their length, probably as a consequence of changes in fetal position and the stresses of parturition. At intervals within the very thick-walled umbilical arteries, there are expanded regions (nodes of Hoboken) with internally projecting ledges of endothelium (folds of Hoboken). Umbilical arteries are unusual in their lack of internal elastic lamina although elastic membranes are present deeper in the vessel wall. Umbilical vein is thinner-walled with a wider lumen and is a low-pressure conduit with

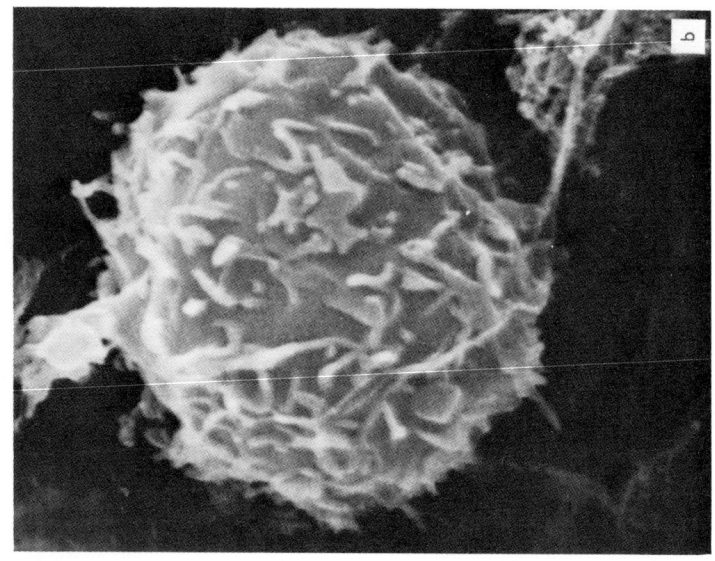

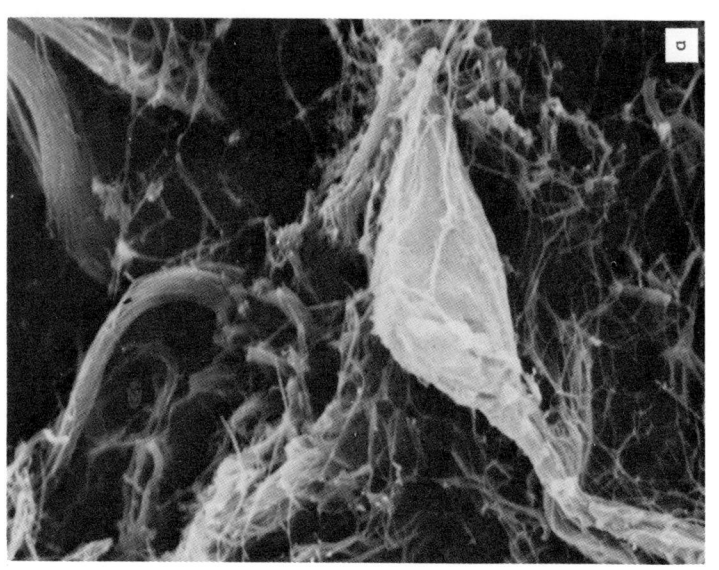

Figure 15.1. (a) Fibroblast in connective tissue of umbilical cord. Note interlacing strands of collagen (X5600). (b) Macrophage attached to collagen network of connective tissues of umbilical cord. Note surface folds and flaps (X11,200) (Hafez et al., 1975; photos by M. I. Barnhart and J. M. Lusher).

blood pressure varying from 8-25 mm Hg after midterm (Cedard, 1971). Thus, there is a great potential for brief periods of interrupted or slow blood flow. Thrombus formation under these circumstances is minimized due to the normal thromboresistant qualities of an intact endothelium (Chen and Barnhart, 1977).

Endothelial cell (EC) patterns differ in umbilical artery and vein although normally there are, in both, tight junctions between cells to limit permeability and possibilities of interaction of subendothelial components with blood components. The organization of EC in the artery displays an X-shaped pattern (Figure 15.2) indicative of some overlapping of surface cell margins and resembles the disposition of endocardial cells within the heart ventricle (Baechler and Barnhart, 1975). Umbilical vein endothelium exhibits EC in orderly and longitudinal arrangement parallel to the direction of blood flow (Figure 15.3).

In response to ischemia (or other stresses), the endothelial cells of the umbilical vessels become quite spindle-shaped and may even release their connections with adjacent cells exposing a forbidden territory (Figure 15.2). Also, EC membrane activity becomes even more vigorous. (Normal activity is prominent, indicative of pinocytosis and exocytosis activities.) Such injury can lead to endothelial cell separation or detachment and exposure of subendothelial fibers and substances that are thrombogenic. Blood platelets may become adherent to the injured vessel wall, forming a pavement as a temporary seal (Figure 15.3). This beneficial hemostatic effect, however, can build into a thrombus that can impede, permanently, the all-important blood flow.

NEONATAL BLOOD

Many differences exist between neonatal and adult blood but the major ones, notable in the blood cells, appear in the erythrocytes (RBC). In the normal newborn, RBC differ from those of the adult in several ways (Barnhart and Lusher, 1975). They are increased in numbers and also in size (macrocytes). Most macrocytes in the neonate appear as large discocytes (Figure 15.4a), although occasional macrospherocytes are seen. Postnatally, RBC decrease in size so that by three months of age RBC

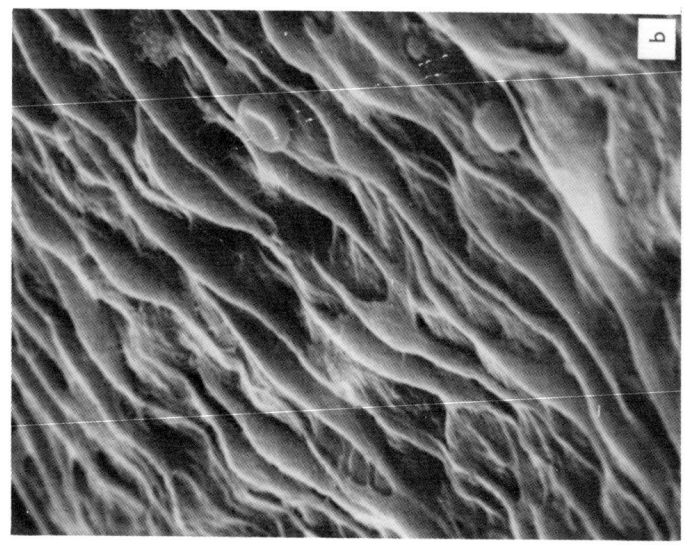
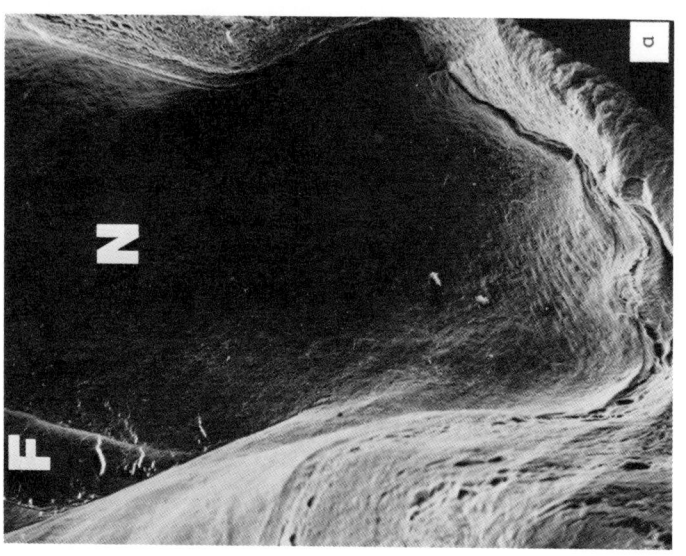

Figure 15.2. (a) Umbilical artery. Edge of fold (F) and node (N) of Hoboken identify this dilated area (X13). (b) Edothelium of umbilical artery. Note that most endothelial cells appear to communicate with one another, producing an X-shaped pattern (X2700).

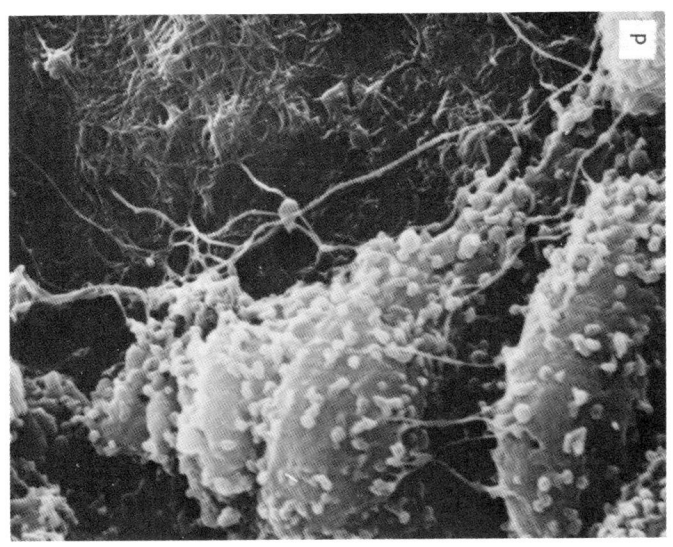
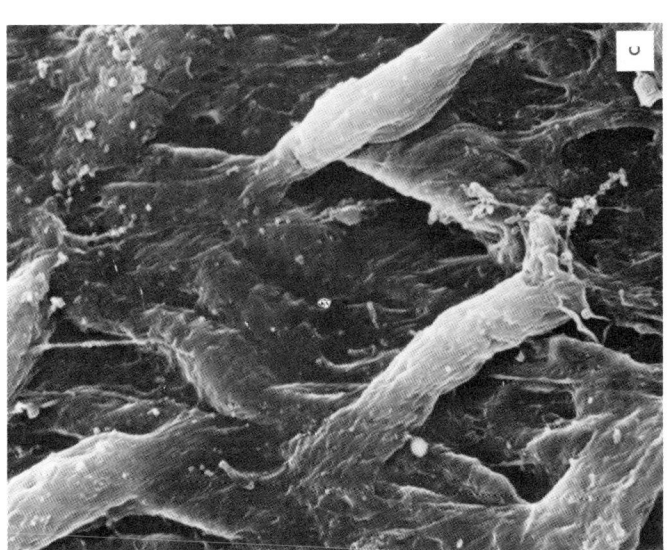

(c) Ischemic endothelium of umbilical artery. Note interendothelial gaps and reduced X-shaped pattern (X1800).
(d) Patchy loss of endothelial cells induced by ischemia and proteolysis exposes the underlying subendothelial fibrillar network. Note increased blebbing of persisting endothelial cells (X4700) (Hafez *et al.*, 1975; photos by M. I. Barnhart and J. M. Lusher).

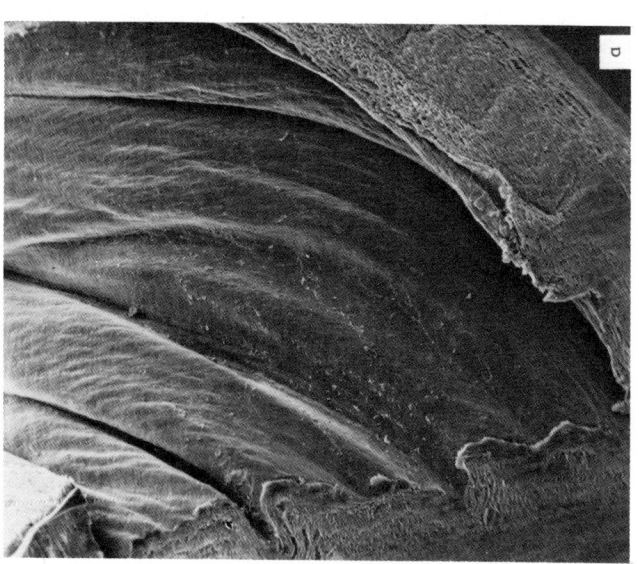
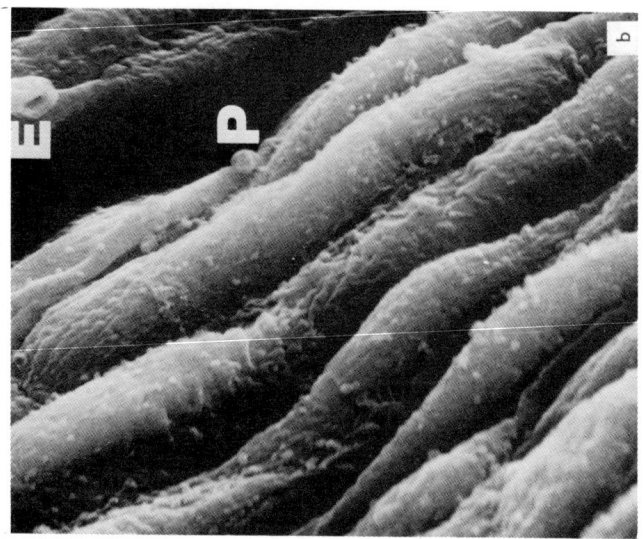

Figure 15.3. (a) Interior of umbilical vein with the broad folds produced by internal elastic lamina (X20). (b) Endothelial cells of umbilical vein lie parallel to one another in the same direction as blood flow. Note surface activity probably signifying pinocytotic activity. An erythrocyte (E) and platelet (P) provide size reference (X2800).

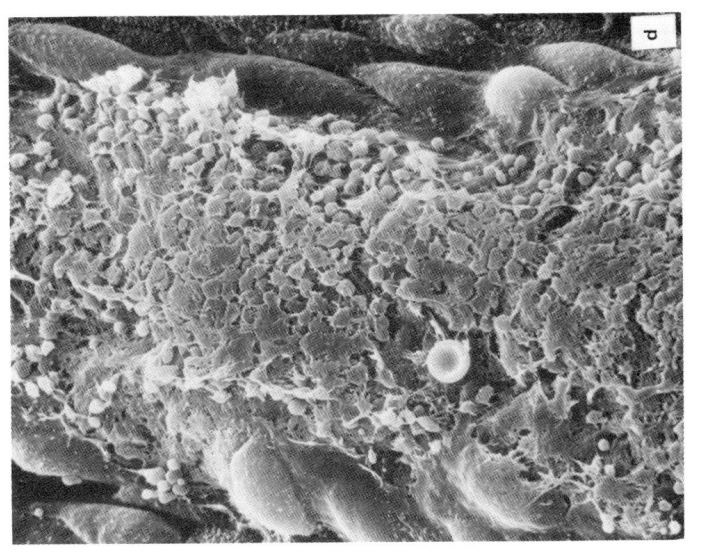

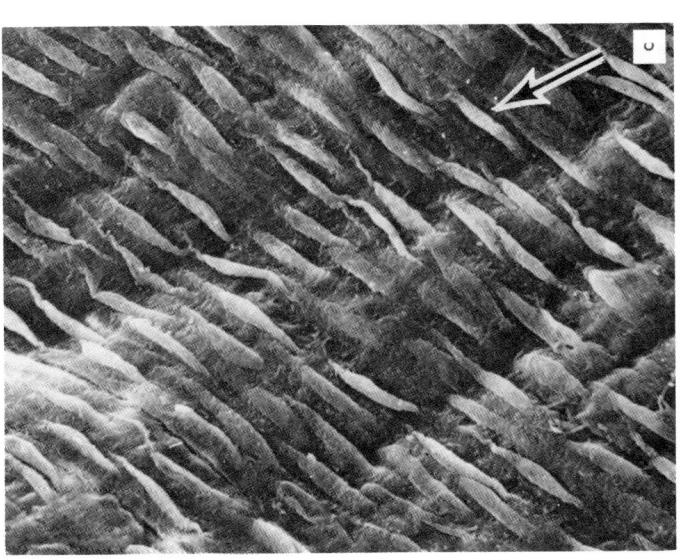

(c) In ischemic umbilical vein the nuclear bulges (arrow) protrude more prominently into the lumen (X5700). (d) Injured endothelium has a pavement of blood platelets as a temporary sealer for vessel wall (X5700) (Hafez et al., 1975; Chen and Barnhart, 1977; photos by M. I. Barnhart and J. M. Lusher).

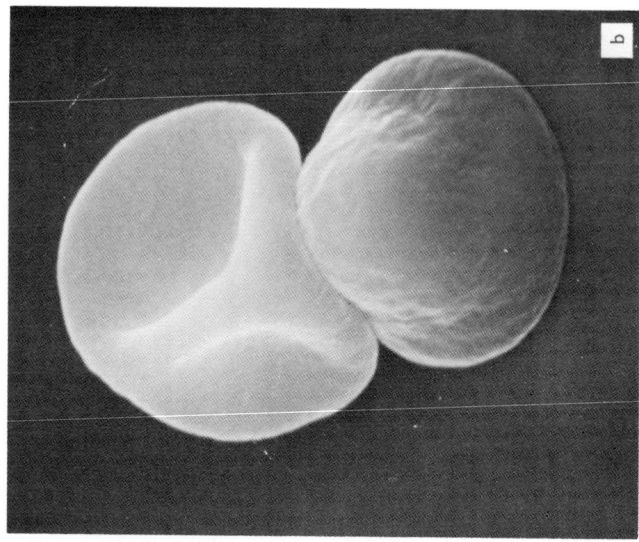
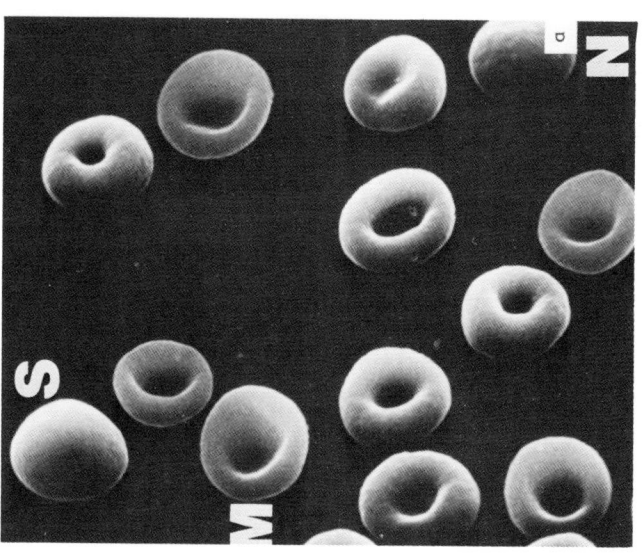

Figure 15.4. (a) Note macrocytes (M) and occasional spherocyte (S) and nucleated (N) erythrocyte typical in blood of newborn (X3000). (b) Nucleated erythrocyte is dome-shaped cell in umbilical cord blood (X10,700).

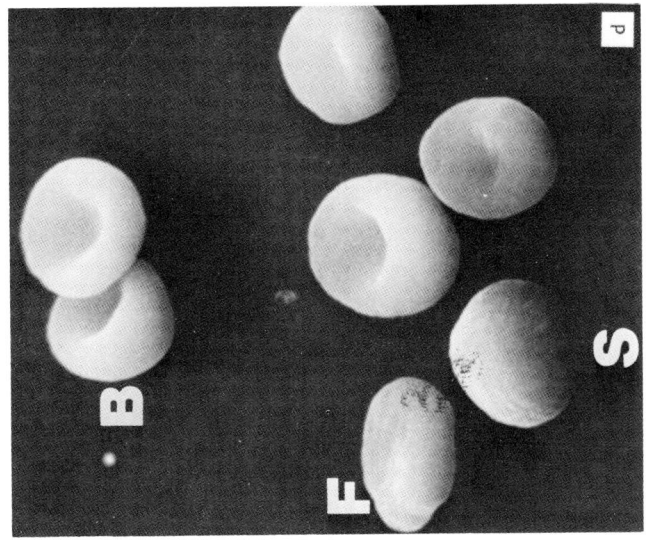

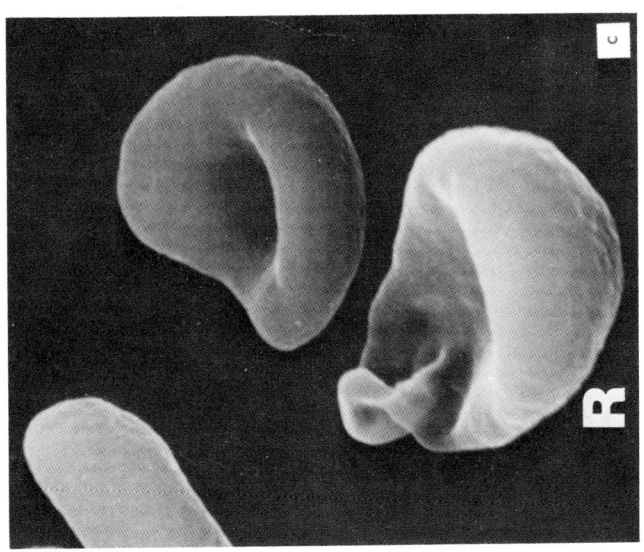

(c) Reticulocyte (R) in blood of newborn (X8250).
(d) Blood of infant with hereditary spherocytosis. Probable membrane abnormality results in bowl-shaped (B), flattened (F) and spherocytic (S) erythrocytes (X4000) (Hafez *et al*, 1975; photos by M. I. Barnhart and J. M. Lusher).

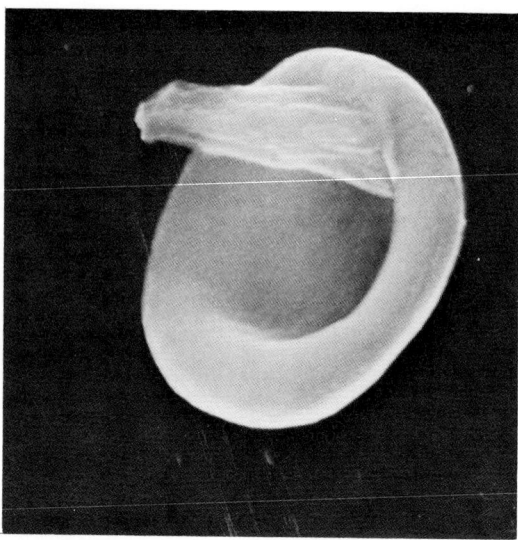

Figure 15.5. Hemolyzing erythrocyte from infant with Rh hemolytic disease. Note extrusion of hemoglobin (X11,000) (Hafez *et al.*, 1975; photo by M. I. Barnhart and J. M. Lusher).

size approaches that of the adult. Further, prominent features in neonatal blood are nucleated RBC and reticulocytes. Their presence reflects the active erythropoiesis which occurs in late gestation and during the first few days of postnatal life (Barnhart and Lusher, 1975). In the normal, full-term newborn, nucleated RBC comprise about 1% of the RBC population but are rare after the third day of life. In the premature infant, the number of nucleated RBC increases. Nucleated RBC are large cells with dome-shaped elevation overlying the nucleus (Figure 15.4b). Reticulocytes comprise 3-7% of the newborn's RBC, but quickly fall off to normal adult values (1%) by one week of age. These young RBC, frequently large, exhibit rather striking surface contours, projections and small pits (Figure 15.4c). Such vigorous surfaces reflect the activity of the reticulocyte as it completes its maturation and extrudes organelles, no longer necessary for its existence as a mature RBC in the circulation.

RBC disorders in the newborn may be either hereditary or acquired. Hereditary spherocytosis is characterized by spherical RBC, which have a shortened survival time in the circulation so that anemia and jaundice may result. These RBC often appear bowl-shaped and display slightly wrinkled surface membranes (Figure 15.4d). In response to the increased rate of destruction of spherocytes, reticulocytosis is frequently observed. Also, the spleen, where trapping and destruction of abnormal RBC occurs, is congested and enlarged.

An example of an acquired RBC disorder is that of Rh hemolytic disease. This condition results from transplacental transfer of maternal antibody directed against the infant's RBC surface membrane antigens. A variable degree of hemolytic anemia results from this interaction of maternal Rh antibody with the infant's RBC. RBC membrane damage may result in hemolysis (Figure 15.5). In addition, reticulocytosis and a striking increase in nucleated RBC are often observed in response to the hemolysis. This is not a pathognomonic sign since it is a compensatory response to various types of hemolytic anemia or hemorrhage when bone marrow reserve and iron supplies are adequate.

REFERENCES

Baechler, C. A. and M. I. Barnhart. "The Umbilical Cord," in *Scanning Electron Microscopy Atlas of Reproductive Physiology*, E. S. E. Hafez, Ed. (Tokyo: Igaku Shoin, 1975).

Barnhart, M. I. and J. M. Lusher. "Neonatal Blood," in *Scanning Electron Microscopy Atlas of Reproductive Physiology*, E. S. E. Hafez, Ed. (Tokyo: Igaku Shoin, 1975).

Cedard, L. "Placental Perfusion *In Vitro*," in *Perfusion Techniques*, A. Diczfalusy, Ed. (Stockholm: Karolinska Institute, 1971), pp. 331-337.

Chen, S. and M. I. Barnhart. "Platelet-Vessel Wall Interaction after Glycohydrolase Treatment," in *Scanning Electron Microscopy/1977*, Vol II, O. Johari, Ed. (Chicago: IIT Research Institute, 1977), pp. 485-492.

Hafez, E. S. E., M. I. Barnhart, H. Ludwig, J. Lusher, I. Joelsson, J. L. Daniel, A. I. Sherman, J. A. Jordan, H. Wolf, W. C. Stewart and F. C. Chretien. "Scanning Electron Microscopy of Human Reproductive Physiology," *Acta Obstet. Gynecol. Scand.* Suppl. 40 (1975).

INDEX

abortus 213,215,216
accessory sex gland 70
acrosomal cap 47,65
acrosome 57-59,64
adinocarcinoma 123
agglutination 65
alcohol 22
alveoli 91,93,193
amniocentesis 177,182
ampulla 107,109
amyl acetate 22,25
androgen 44
azoospermic 73

basement membrane 45,50
blood
   neonatal 217
   platelets 219,223
   -testis barrier 42,51
boundary tissue 41
breast 193,195-197

carcinoma 13
   in-situ 155,156
   Also see adenocarcinoma
cells
   amniotic fluid 177
   associations 44,51

basal 92
ciliated 4,6-9,12,107-109,116,
   119,134-137,139-144,160
   columnar 148,150
cuboidal 99
endothelial 219-222
germ 39,42,43,45
germinal 46
glandular epithelial 91
granulosa 101,103
kinociliated 118
metaplastic 154
microvillous 211
myoepithelial 201
nonciliated secretory 4,5,7-9,
   12,107,113,135-137,139-144
   columnar 148-150
"peg" 135,139
peritubular (Myoid) 41,42
sebaceous fat 13
secretory 119
Sertoli 40-43,47,49-53
stem 45
types 4
   pathological 120
cervical
   crypts 159,160
   crystallization 163,168,170,171
   microfibrillar 165
   rheological characteristics 159

cervix   127,147-150,152,153
    uteri   147
    uterine   22
    Also see endocervix
chorion   215,216
chromosomes   45
cilia   107,128,134
    solitary   100
    Also see ciliated cells, stereocilia
ciliary basal bodies   12
ciliogenesis   116
clone   45,46,49,53
coagulating gland   71
coagulation   58,69
coagulum   74,80,82,84,86
    formation   71
collagen   217,218
concretions   94,95
conducting   25
cork plates   19
Corpora amylacea   92
cortex   99
crown-rump length   204
corona radiata   101,102
cumulus oophorus   101
cytoplasmic
    bridges   45,209
    droplet   47

dehydration   22
desmosomes   49,52
Diethylstilbestrol   13
drying   22,25

ectoderm   203,208-210,212-214
ejaculate   57,59
    split   71
ejaculation   70
embryo   203
    development   205,213

41-day-old   212
six-week-old   204
embryonic period   203
endocervix   116
endoderm   203
endometrial glands   5,134
endometrium   116,128,135,136,
        140,141
    cornual   134,137,138
    pathology   123-127
endoplasmic reticulum   182
endosalpinx   134,135,139,140,143
epididymis   44
epithelium   196
    alveolar   196
    duct   196
    mature squamous   152
    metaplastic squamous   147,151
    original columnar   147,149,151
    original squamous   147
    renal   13
    Also see vaginal epithelium
erythrocytes   219,224-226
estrogens   141,171
evaporation   25

fallopian tube   133
ferning   164,171,172
fetal
    maturity   3
    period   203
    prematurity   13
fetus   203
    development   213
    nine-week-old   208
    61-day-old   212
fibers   74
fibroblasts   217
fimbriae   102,107,108
fixation   22
Freon   22

glutaraldehyde 22
glycogen 183
glycoproteins 129,161
Golgi complex 182
gonocytes 44,45

hyperplasia
 endometrial 127

infertility 13
infundibulum 141
intercellular
 bridges 211-213
intracellular junctions 52,53,183
ischemia 219,221,223
isthmus 107,109

kinocilia 8,9,12,101,129
 Also see kinociated cells
Krebs-Ringer glucose 22

lactation 193
lactiferous duct 193
leukocytes 64
liquefaction 58,69
liquid carbon dioxide 22
lobes 193
luminal fluids 12,40

macrocytes 219,224
macrophages 217,218
mammary glands 193
meiosis 39,45,46
mesoderm 203
metal coating 25
metaplasia 151
 Also see metaplastic cells

micelles 160,161,163
microfibrils 159,161-163,167
microplicae 183,185,187,188
microvilli 4,7,8,99,116,118,124,
 126,129,134-136,139,142,
 148,183,187,188,199,201,
 209,212
 Also see microvillous cells
midpiece
 abnormalities of 64
mitochondria 182
mitotic arrest 45
mounting 25
mucins 161,173
mucus 22,129
myoepithelium 196
myoma 125,126

Nomarski optics 14
normospermic 73

oligospermic 71
organogenesis 204
ovary 99
 fetal 100
oviduct 107
ovulation 12,103,171,173
ovulatory follicle 101
ovum 101

pathological
 changes 121
 Also see cell types
perfortiosum 65
phases
 follicular 171
 luteal 162,171,173
 ovulatory 111
 proliferative 111,134,163
 secretory 111,134,135

physiological saline   22
pinning   19
plasma
  -lemma   64,88
  membrane   64,183,187
  seminal   65
pneumocytes   13
polyp
  endometrial   127
postmenopausal period   140
postmenopause   142,143
postnuclear cap   64
postovulatory
  follicle   103
  period   134
preovulatory follicle   99
progesterone   141,171
prostate   71,89
  parenchyma   91,94
  stroma   90
puberty   40,44

reticulocytes   225,226
refractile fibers   73
reproductive tract   4
residual bodies   47
resting gland
  parenchyma   196
  stroma   196
rugae   10

saccular recesses   91,93
salpingitis   12
secretion   92
secretory
  activity   113
  functions   129
  materials   116,117,128
  Also see secretory cells
semen   57,69

seminal
  fluid   89
  vesicles   71
seminiferous
  epithelium   44
  tubules   40-42,44,45,50-53
seminin   73
species differences   69
sperm   9,45
  fibrin   73
  head   58,64
  tail   63
spermatids   39,43,46,48-50,51,53
spermatocytes   43,46,50,52
spermatogenesis   39
spermatogonia   39,41,43-46,49,50
spermatozoa   13,39,57,70,80,81, 117
  abnormal   59,61
  phagocytosis of   63,64
spermiation   39,42,45,47
spermiogenesis   42,44,46,47,49, 52,53
spherocytes   227
sputtering   25
squamo-columnar junction   148
sterocilia   99
sterility   13
stigma   101,102
stroke
  effective   9
  recovery   9

testis   48-50,57
  mature   41
  rete   41,44
  Also see blood-testis barrier
testosterone   39
theca interna   99
tight junctions   219
tissue organization   4

transmission electron microscopy
    13
tumors 89
tunica albuginea 99,103

umbilical
    arteries 217,220,221
    cord, 217
    vein 217,219,222,223
uterocervical junction 6,7
uterotubal junction 133

uterus 12,107,147,148

vagina 148
vaginal
    epithelium 10
    plug 70
variocele 59
vesiculase 71
villi 216

wave concept 44